More Than Words

More Than Words

How Talking *Sharpens* the Mind and *Shapes* Our World

Maryellen MacDonald, PhD

AVERY
an imprint of Penguin Random House
New York

an imprint of Penguin Random House LLC
1745 Broadway, New York, NY 10019
penguinrandomhouse.com

Book design by Daniel Brount

Library of Congress Cataloging-in-Publication Data

Names: MacDonald, Maryellen Coles, 1960– author.
Title: More than words: how talking sharpens the mind and shapes our world / Maryellen MacDonald.
Description: New York: Avery, [2025] | Includes index.
Identifiers: LCCN 2024041741 (print) | LCCN 2024041742 (ebook) |
ISBN 9780593545270 (hardcover) | ISBN 9780593545294 (epub)
Subjects: LCSH: Speech. | Psycholinguistics. | Psychology.
Classification: LCC BF455 .M214 2025 (print) | LCC BF455 (ebook) |
DDC 401/.9—dc23/eng/20250212
LC record available at https://lccn.loc.gov/2024041741
LC ebook record available at https://lccn.loc.gov/2024041742

Printed in the United States of America
1st Printing

The authorized representative in the EU for product safety and compliance is Penguin Random House Ireland, Morrison Chambers, 32 Nassau Street, Dublin D02 YH68, Ireland, https://eu-contact.penguin.ie.

For Mark
and for Claudia and Ethan,
my very favorite talkers

CONTENTS

Introduction

"I'M WRITING A BOOK ABOUT TALKING, BUT IT'S NOT ABOUT communication." That's what I would say to people who asked what my book was going to be about. I quickly learned to sneak "it's not about communication" into the very first sentence of my answer. If I didn't, folks would tell me about their favorite aspect of communicating and ask how I was going to cover it in my book. Maybe I would have advice about how to talk to teenagers? Was I writing about how men and women communicate differently? Was I saying that politics is now so polarized that people can't talk to each other anymore? There were some crestfallen expressions and even pushback when I said no, the book is not about how we communicate. One person quite vehemently told me that I should toss out whatever it was I was writing and instead produce a self-help book about how to talk better.

All of these conversations made me more determined than ever to write about the hidden noncommunication side of talking. Everyone knows that we talk to communicate. There's no shortage of books, podcasts, and advice columns about communicating with each other. Many of them are quite excellent. I didn't have a calling to add to this pile.

But I did see a real need to tell a different story. It's about a side of talking that most people have never heard of, but which nonetheless has profound effects on all of our lives. We humans have the capacity to transform our internal ideas into something that others can perceive—speech, writing, or signing in a sign language. And sometime during childhood, most of us also begin to talk to ourselves, an internal monologue that others can't hear or see. I'm going to use *talking* to refer to all of these ways that we produce language—speech, sign language, writing, and internal speaking or signing to ourselves. They're not identical, but they share a key feature that is important for the theme of this book: All of these idea-to-talk transformations require serious mental effort, and our brains have figured out how to streamline the workflow of talking. The brain's strategies are brilliant and fascinating, but here we get to a twist in this story.

The ways that the brain solves the challenges of talking don't just let us communicate; they also have consequences that have almost nothing to do with communication. The brain's talking routines spill over into other brain processes, changing our lives in surprising ways. Most of these side effects of our talking skills are overwhelmingly positive for us. And they're really hidden: Even scientists who study them haven't necessarily linked them to our ability to talk. That's the story I wanted to tell.

The noncommunicative effects of talking are a varied bunch, studied by different scientists who don't frequently interact with each other or even know that much about talking. Some of these effects show up when infants are just learning to talk; others support our final years. Most are with us throughout our lives. Here's a sample:

- Within the first months of life, babies' silent pre-talking tongue movements sharpen their ability to perceive speech and learn the language around them.
- From childhood onward, our talking focuses our attention and helps us stick to our goals.
- Talking, even internally to ourselves, modulates intense emotions and helps us cope with upsetting events.
- Talking helps us maintain our memory and may prevent or delay dementia.

And that's just the start.

If these benefits seem unrelated to one another and unrelated to talking, well, that's really my point. Focusing attention, settling emotions, and perceiving speech may seem like a random grab bag of skills because you don't know about how our talking machinery works. When I tell you about what we do in order to talk, these seemingly unrelated facts will fall into place, with talking as the glue that holds them together. In some cases, you'll see how you can be more strategic in your own talking to reap these benefits more fully.

Maybe it sounds a little odd to say that I'll explain how talking works, because of course you already know how to talk. But what you don't know is how you do it. Don't feel bad; you don't know how you talk because your brain won't let you know.

Here, I'll show you what your brain does and doesn't allow you to know about talking. How do you order something at a food counter, like a cup of coffee or a sandwich? You might say that you look over the menu and then say something like, "I'll take a small coffee, uh, and maybe one of those cookies." Perfect, good answer, because it reveals the difference between what your brain does and doesn't let you become aware of. You consciously know certain things, like that you should approach the counter, wait your turn if necessary, settle on what you want. Maybe later you also realize you shouldn't have ordered that mealy cookie.

You navigate the conversation at the food counter with ease, but you're clueless about how you do the actual talking part. How do you convert a desire for coffee into making appropriate words come out? It's just a mystery: When you want to communicate something, somehow your brain makes it so.

The brain doesn't give up its secrets easily, but scientists have figured out a lot of what's going on behind the scenes. We'll see how the brain converts an internal idea or desire into talk that others can perceive. That conversion process is fascinating all by itself, but the real payoff in this book is understanding how the idea-to-talk conversion process has important noncommunicative consequences: Because the brain uses a particular routine for picking words to say, this word-picking process focuses our attention. Because these word-picking processes reflect our ideas and attitudes, analyses of our talk patterns can reveal aspects of our thinking, personality, and mental state. Because the brain uses certain procedures to arrange words in order, talking changes our language over time. Because talking is hard mental work, it helps us stay sharp in old age.

When you understand the links between talking and its consequences, you'll get a deeper understanding of remarkable interconnections in our brains, linking situations as disparate as babbling babies and sports psychology. You'll understand what infants' *babababa* babbling is doing for them, and what you can do in return to help your baby. You'll learn the talking-based reasons why phones, tablets, and other media for infants and young children should be limited, and why you should limit your own media use around kids. You'll understand the myriad advantages of talking to yourself—really, you can stop criticizing yourself about it—and you'll also become more alert to potential downsides of this habit. You'll know how to deploy talking to boost your learning, and how to help kids, students, and anyone else use their own talking to learn and remember better.

I MAY BE SPINNING SOME REVISIONIST HISTORY HERE, BUT I BElieve I've been thinking about talking since I was a little kid. My first memory of the power of talking is from age five. My parents, who'd spent their entire lives in Boston and New York, picked up and moved us to a small town in Texas, sight unseen. As we were getting ready to move, my father explained that people would talk more slowly in Texas. He tried to illustrate by first saying a "regular" example of the sentence "Hello, little girl" in his Boston accent, followed by what he imagined to be the Texas version. It was still in his Boston accent, but slower.

When we got to Texas, I wasn't struck by people talking slowly; instead, I was overwhelmed by the utter *differentness* of everything coming out of people's mouths. Different sounds, different words for things, different manners, different expectations about how and when children should talk, and much more.

On the first day of kindergarten, I asked where the bathroom was and was told that I wasn't a good listener. Earlier, the teacher had pointed out the location of what she called the restroom. I had never heard that word before, and I assumed that the restroom was the room for resting. Would they make us take naps in kindergarten? But no, a restroom turned out to be a bathroom by another name. This was the first of many experiences where I thought I knew what was going on, but then a new word would show up to upend that feeling. It was confusing, but it also made me think.

That same year or maybe the next one, I noticed that my very favorite Crayola crayon, violet blue (a lovely shade now inexplicably "retired" by the Crayola company) had a cousin of sorts in the crayon box, blue violet. And there were other pairs of cousins: green yellow and yellow green, red orange and orange red. I spent a rather large amount of time puzzling about the fact that the second word in these crayon names specified the color—violet blue is a kind of blue—and the first word added some sort of flavor to it—violet blue is a blue with a bit of violet in it. Why do we say the words in that order and not the other way around?

These puzzles, along with wondering about the different accents I kept encountering, led me to find languages and talking deeply interesting. I even got a PhD in the field of psycholinguistics, straddling the psychology and linguistics departments, focused on how we use language. My education and professional life kept me moving from one accent to another. I began at the University of Texas at Austin; had two stints in Los Angeles, at UCLA and USC; spent some time in Pittsburgh at Carnegie Mellon University; made several stops in the Boston area at Northeastern University and MIT; and finally moved to the University of Wisconsin–Madison. Each place exposed me not only to new knowledge and research about how we use language but also to different accents, words, ways of talking.

Like most other psycholinguists, I began my career studying how we comprehend language. But I was always a little restless, and I kept finding other topics to study. I first branched out into studying language comprehension in elderly adults and those with Alzheimer's dementia. I also dabbled in how adults learn second languages and how children learn to read. And I did a deep dive into the relationship between our language abilities and our short-term memories.

These are all very interesting topics, but I didn't stop there. Eventually I started studying how we talk. I was shocked to realize just how small this field of study is. Psycholinguistics itself is a rather niche area, and talking research is a tiny corner within it. The more I learned about how talking works, the more I thought that the field was unjustifiably overlooked. I was repeatedly struck by how the mental processes behind talking led to key insights into other areas I'd studied and puzzled over. I wrote scholarly papers about how the inner workings of talking help explain quite a bit about how we comprehend language and how our short-term memories work.

Each fall when I taught a psycholinguistics course to freshmen and sophomores, I kept working in more and more material on talking. Students found it surprising and interesting, especially when I described how talking had useful side effects that spill over into other aspects of their lives. These benefits of talking kept piling up as

I taught and did research, but no one seemed to be putting them all together. Eventually I realized that there was really a story to tell here, and that I, who'd worked in many relevant areas, could be the person to tell it.

In telling that story, I've divided the book into three parts in a way that mixes information about how you talk—the secrets that your brain doesn't let you know—with the side effects that emerge from what your brain is doing when it's creating talk. Throughout the book I'm going to use *talking* to mean any internal or overt form of producing language—speaking, writing, signing in a sign language. Sometimes we'll make distinctions among these forms, but usually the nature of talking and its benefits are the same no matter which form you talk in.

Part 1 is about what talking is, who can do it, and how it works. We'll observe interspecies communication in the African rainforest and check in with little babies who are on the on-ramp to talking. We'll see that while many nonhuman species are shockingly smart and can understand some amount of human language, only humans can talk. That means that only humans reap the benefits of talking that are the core of this book.

Why this strict divide between the talkers and non-talkers of the animal kingdom? There are a cluster of reasons, but one big one is that talking is massively more difficult than comprehending someone else's talk. The special difficulty of talking, and how the brain develops efficient strategies to cope with the difficulty, are essential to creating many of the valuable side effects of talking.

We'll see many of these by-products of talking in part 2, which homes in on the benefits of talking for our own brains. Because of the way talking works, our attention, our motivation, and our learning all get a boost from talking. That's excellent news, but our society isn't necessarily set up to let us fully realize these benefits. For example, the ways that talking enhances learning could revolutionize teaching, but any change runs up against an educational establishment that actively discourages talking in school. Similarly, talking has proven benefits for helping elderly adults stay sharp and independent as they age. Promoting more talking might lower spiraling health care costs and improve the lives of seniors and their families, but there's no push to take advantage of the talking benefits for older adults. I'll offer suggestions for how we might chart a different path.

Part 3 digs into how our talking affects aspects of the world around us. We'll look at large patterns across the world's languages and ask how they're shaped by the way talking works. We'll also examine the social consequences of variations in the way people talk. Even when everyone's using the same language, the exact way someone talks reflects their age, class, race, and the region where they grew up. This local development of dialects is a completely natural and unavoidable consequence of how we learn to talk, and yet there's a huge backlash against different ways of talking. To put it mildly, many folks don't like the fact that other people don't talk like they do. This resistance is itself a natural consequence of how we learn to talk. Perhaps if folks know a bit more about how talking works, they might let go of some of these prejudices.

Another natural consequence of the brain processes that create our talk is that talking patterns are unique to each individual. We'll discuss how computer-aided analyses of a single person's talk can help predict whether that person is likely to need medical care, commit a crime, or want to buy a car. All of this might sound pretty futuristic, but it's becoming increasingly common. Companies know that your talk patterns are a valuable commodity, and you should know about this too.

When I first thought about writing this book, I wanted to use my own expertise as a language comprehension researcher to put myself in the reader's position. I'd spent years studying why some kinds of language are easier to comprehend than others, and I was determined to use that expertise to try to write a book that many people would enjoy reading. I realized that one big difference among science books for general audiences was in how the author described the scientific research. In novels, biographies, and memoirs, readers get to know the characters, care about them, and can keep track of them. Science books work differently. The characters—the scientists—sometimes pop up out of nowhere when an author mentions them in connection with some research, and then that character disappears forever or returns in a completely different context five chapters later. As a result, general audiences have no motivation to remember or care about the effectively anonymous researchers who make cameos throughout a book; the names just add clutter for the reader who's really interested in the ideas and their implications.

For these reasons, I decided I would name the scientists behind the research only in the endnotes and not in the main text of the book. This was a really tough choice for me, because it goes against best scientific practices of giving credit to those who have advanced our knowledge. I also know many of these researchers as students, friends, and colleagues, and no friend wants to be ignored. Nonetheless, my thinking about language comprehension suggests that omitting most names will create a more engaging and accessible narrative.

I occasionally make an exception to this rule when I think that mentioning someone improves the flow of the book, for example when I interview a researcher or when I need to refer to them repeatedly. Whether they get a mention in the main text or not, I have listed researchers and their popular and scholarly works in the endnotes, both to give credit and to provide resources for those who want more information.

What I do include in this book is the remarkable story about how the process of talking, which your brain doesn't let you know about, becomes the key to understanding so many qualities that make us human. The story also has a classic conflict between science and society, where our attitudes and social institutions don't understand the nature of talking and therefore do not allow us to fully reap its benefits. Even though this book isn't about communication, I think I can communicate all of this interesting stuff to you. Let's get to it.

PART ONE

Getting to Talk

CHAPTER 1

We Do the Talking

"I have been following you, my son," said the chickadee, "and I remember that you insulted me once, but that you were very sorry. You fed me. You asked my pardon. Therefore I will help you now."

LOUISE ERDRICH, *CHICKADEE*

I LOVED THE SERIES OF DOCTOR DOLITTLE BOOKS AS A KID, where an introverted physician gave up his medical practice and learned to speak and understand the languages of the animals around him. He and his animal friends had amusing cross-species conversations and went on fantastic adventures. Most of the people he encountered thought he was an utter crank. The hidden animal society Dr. Dolittle inhabited, the secrets the animals told him, and the individuality of the animal characters were all fascinating to me and to millions of other children.

Dr. Dolittle is not alone. So many myths and stories, spanning centuries and cultures, contain animals who have important things to say to us—guidance for the hero, prophecies for the future. If this literary theme can be said to reflect our hopes and desires, then all those talking animals are showing us something important about ourselves: We dearly wish that animals could talk to us.

Alas, we're out of luck. In the real world, animals don't talk. Some animals do understand a portion of what we say to them, but they don't talk back. The absence of animal talking while understanding some talk can teach us a lot about talking and what it means for us. We humans reap incredible gifts from talking, going way beyond the obvious benefit of communicating with one another. These extra gifts are unavailable to all the non-talking species, even those that can comprehend some language.

In this chapter, we'll explore how animals that don't talk nonetheless communicate with each other. We'll look at communities of monkeys that use alarm calls to warn each other of predators. We'll see how other species are listening in on the monkeys' communication. Out in the wild, understanding the signals of another species is alive and well, and quite beneficial to the eavesdroppers.

We'll then consider whether we can expand cross-species communication and train other animals to understand our human talking. Spoiler here: maybe so! I'll describe apes and dogs that have learned hundreds of human words, suggesting that the multispecies communication in the wild can be extended to at least some understanding of human talking.

We'll also discuss what these language-trained animals say to humans in return, and what they don't say. The comparison between animals' ability to understand human talk and their ability to talk themselves turns out to be very telling. It will set us on the path to thinking more deeply about talking—what it is, what's special about it, and how talking changes us humans in so many ways.

Animal Communication and Eavesdropping

A Diana monkey sits high in a tree in a West African rainforest. She's quite a beauty. Diana monkeys were named for the Roman goddess Diana, often portrayed with a prominent brow and an intense stare. Our monkey is definitely rocking that look. She has a fierce jet-black face, crowned by a white unibrow and matching ruff. When she yawns, she shows off fangs that a vampire would envy. Her mate has all the above and an additional gorgeous accessory that gives new meaning to the term *family jewels*—his scrotum is a dusky turquoise blue.

Many animals, including Diana monkeys, make calls that signal the caller's emotional state or social status in the group. Diana monkeys also have something more unusual in their vocal repertoire. They have calls that identify specific predators in their environment.

Diana monkeys make a loud call when they see an African crowned eagle. These enormous birds are fierce hunters, strong enough to kill a small antelope. They hunt monkeys by surprise attack, swooping into the trees to grab an unsuspecting target. When one monkey gives the *eagle* call, the others search the sky and give more alarm calls. If the adult male in the troop spots the eagle in the sky, he moves to a location where he can better keep an eye on it. If he sees that the bird is in the monkey troop's own tree, the male shows that his turquoise blue cojones are made of steel—he bares his fangs and races toward the fearsome eagle to attack it.

Diana monkeys have a different call for another stealth attacker, leopards. Their eagle and leopard predator calls, called *hacks*, sound a bit like a bark, with distinct overtones of dredging up phlegm. To untrained human ears, the *eagle* and *leopard* calls sound identical. The primate communication expert Klaus Zuberbühler told me that when he first heard these two calls in the jungles of West Africa, he thought that the Diana monkeys had only one generic predator call that covered both eagles and leopards. At that time, Klaus was a novice researcher, observing primates in the wild for the very first time. He'd been sent to Côte d'Ivoire by senior scientists who had heard that Diana monkeys might be doing something interesting with their calls. The researchers wanted someone to go check out the situation, and Klaus got the assignment.

Klaus's initial impression that the monkeys had only one general predator call was extremely disappointing. Many species have a generic alarm call, and it seemed like there wasn't anything special about the Diana monkeys' vocalizations after all.

Klaus, who's now a senior scientist himself, told me that this first project with the Diana monkeys was tough to get going. The Diana monkeys were terrified whenever they noticed a human nearby, and they ceased all calls whenever Klaus or other researchers tried to observe them. For a project studying monkey calls, it was a major setback when all the animals would fall silent whenever it was time to collect some data. Eventually, slowly, Klaus and his colleagues figured out ways to observe the monkeys without being detected, and we now know what the Diana monkeys knew all along: There are two distinct hacking calls, one for eagles and another for leopards. The monkeys have no trouble telling the two calls apart, and now Klaus can easily hear the difference too.

In the process of studying the Diana monkeys, Klaus and colleagues discovered something else. In the Diana monkey neighborhood, other species were listening in on the Diana monkeys' *eagle* and *leopard* calls. And Diana monkeys were eavesdropping on other species too.

The realization that different species were listening to each other began with studies homing in on how the Diana monkeys' calls worked. In one experiment, Klaus and colleagues hid a loudspeaker in the underbrush in the territory of a Diana monkey troop. Then they played a recording of a Diana monkey making the *leopard* alarm call. There was no sight, smell, or sound of a leopard anywhere about, but hearing the recorded call, the Diana monkeys looked down and searched the ground. Their behavior showed that the monkeys do understand the call to predict the presence of a leopard.

Soon Klaus and other researchers noticed that other tree dwellers were responding to the alarm-call recordings. Campbell's monkeys, a nondescript species that shares the trees with Diana monkeys, have their own *leopard* and *eagle* calls. Campbell's monkey calls don't sound like the Diana monkey calls, but Diana monkeys and Campbell's monkeys respond to each other's calls in the same ways that they respond to the calls of their own kind.

The cross-species understanding doesn't stop with monkeys. Yellow-casqued hornbills are large birds that forage in the trees near the Diana and Campbell's monkeys. They rival Diana monkeys in their striking appearance. They're mostly black, with a yellow beak almost as long as their entire body, and a bedraggled Mohawk of feathers on their heads. They peer around with googly eyes that look like they came from a kid's craft project.

The hornbills have no special predator calls themselves, but this doesn't stop them from tuning in to monkey calls in their environment. Like the monkeys, hornbills are subject to sneak attacks by crowned eagles. When a group of hornbills hears a monkey's *eagle* call, they scan the sky with those googly eyes. If they spot the eagle, the whole flock flies out screaming, attempting to drive the eagle away. The hornbills and monkeys have a mutual aid society going—the monkeys can alert the hornbills to the presence of eagles, and the birds can harass the eagle in the air and spoil its sneak attack.

To the hornbills, the very similar-sounding *leopard* call is a different story. When a monkey makes a *leopard* call, the hornbills all . . . do nothing. They live so high in the trees that leopards can't prey on them. While the Diana and Campbell's monkeys are all on alert and scrambling around looking for the leopard, the birds are literally above it all.

We can draw an important lesson from the hornbills going on alert after *eagle* calls while ignoring *leopard* calls: Cross-species attention to calls happens only when it's beneficial to the animal perceiving them. If another species' communication is not directly relevant to an animal's survival, it is ignored.

There is one more group that can benefit from paying attention to *leopard* and *eagle* calls—the predators themselves. Both eagles and leopards need surprise to catch a monkey, and if Diana or Campbell's monkeys are calling out about having just spotted them, it makes sense for these predators to try their luck elsewhere. And that's what they do. Leopards are more likely to leave an area after hearing a monkey's *leopard* call, and eagles are more likely to leave after hearing a monkey's *eagle* call.

So much of this activity in the rainforest—hornbills mobbing eagles, monkeys repositioning themselves in the trees, leopards leaving the area—is prompted by some animals calling and others getting information from those signals. These complex interactions around a single tree are not just fun facts from the jungle; they show that species naturally eavesdrop on each other and use what they hear. Can we extend this natural cross-species communication to deliberately train animals to understand humans? That's where we're going next.

Can Animals Understand Our Talk?

Getting an animal to understand even a bit of human talking likely requires a smart animal, lots of training, and at least one more crucial ingredient: motivation. Just as the hornbills ignore the monkey calls that don't do them any good, we'd expect other species to ignore us if there's not an extremely strong reason to pay attention to human communications.

Many research projects have tried to create motivation and training regimens in order to investigate whether animals can be trained to understand some of what we say. There's a long list of species that have been studied, including dogs, chimpanzees, close relatives of chimpanzees called bonobos, orangutans, gorillas, dolphins, parrots, and parakeets. The list of ways that humans do the talking in these experiments is also long, including speaking English or another human language to the other species, signing in a system modeled on human sign languages, other artificial gesture systems, and underwater signs and sounds for dolphins.

Dogs, maybe even yours, are a great example of a species that can extract some meaning from human talk. Most dogs can learn to attend to their name and some spoken and gesture commands. Border collies, who are bred to pay close attention and respond to spoken, whistled, and gestured commands from their humans, appear to be the champion cross-species understanders, with several of them trained to respond to hundreds of spoken words.

A border collie named Chaser holds the record for understanding human speech. She was raised by a professor of animal cognition who trained her for hours every day over many years to respond to spoken English. Chaser eventually learned the names of more than one thousand different toys. The endnotes to this chapter have a link to a YouTube video of Chaser being put through a challenging test, where she's asked to retrieve toys that a human names.

The massive difficulty of remembering the names of all those toys is rather amusingly illustrated by the behavior of Chaser's human trainer: He marked each of Chaser's toys with its name so that he could remember and say it correctly when he was training her. Chaser, whose abundant skills did not include reading, couldn't benefit from the name printed on the toy. She had to remember which toy was called Blue, and which was Dolphin, and SantaClaus, Bear, FratRat, SpongeBob, Inky, Prancer, Croc, and on and on for 1,022 different toys. Even though there were thousands of hours of training behind Chaser's ability to link words to toys, it's still an astonishing feat of memory and cross-species word understanding.

Although Chaser holds the record for word understanding by a nonhuman, some apes also appear to be able to distinguish hundreds of words in human speech. The clearest examples of ape understanding come from studies of chimpanzees and bonobos that were conducted at the Language Research Center of Georgia State University in the 1980s.

The researchers intensively trained the apes to relate spoken words to nonsense symbols on a giant keyboard. When an ape touched a symbol on the board, a computer played a recording of the word that was associated with that symbol. The humans spoke to the chimps and bonobos in English, and the apes could demonstrate their understanding by touching the symbol associated with the word that the humans had spoken. In a video linked in the endnotes, a bonobo named Kanzi sits in front of the board, touching symbols when one of his handlers says a word. Kanzi learned to touch the correct symbols for about 360 words.

Whereas Chaser the dog's training focused on her huge pile of toys, many of Kanzi's words were related to foods and beverages that he liked, including *milk, juice, onions, apple,* and *banana*. This difference in word choices for dogs and apes isn't accidental; it reflects different species' motivations to learn. Animals, whether they're wild hornbills, pet dogs, or apes in language experiments, do the hard work of understanding another species only when it benefits them. Chaser's border collie herding instincts and desire to please her human can be extended to have her follow commands to round up her herd of toys. Bonobos are not domestic animals, and they have no instincts to herd anything or to please humans. What they do have is strong food preferences. Trainers can use the apes' desire for tasty food to motivate their learning of food words.

Kanzi was asked to follow commands to test whether his understanding extended beyond simple associations between a word and a key on his keyboard. In these tests, Kanzi sat near an array of objects, such as a ball, some juice, a stick, and so on. One of his trainers would instruct him to perform an action with some of these objects, and Kanzi would attempt to follow the request. Some of the instructions were very unusual, designed to test Kanzi's understanding of exactly what he heard, rather than just guessing what his trainers probably wanted. An example of an unusual command is "Put some soap on your ball." Kanzi did fairly well following the commands. At about eight years into his intensive daily training, his ability to follow commands was rated as similar to the capabilities of a two-year-old human child.

Claims of humanlike abilities and the complex role of animal motivations crop up in a short story written more than fifty years before the Kanzi project began. In Franz Kafka's "A Report for an Academy," a wild ape was shot and captured in Africa. He was sent to Europe, given the name of Red Peter, and taught the ways of human beings, including language. In the story, news of Red Peter's abilities goes viral. He's a celebrity in high demand everywhere; even a learned scientific society wants to hear from him. In his report to the scientific academy, Red Peter is quick to dispense with all the warm and fuzzy animal-human bonding myths that are swirling around him. "Imitating human beings was not something which pleased me," he told the scientists, "I imitated them because I was looking for a way out, for no other reason."

It's clear that Kafka's onto something here. He captures the insight that animals will attend to our ways only when it is in their own strong interest. He conveys the animals' ambivalence to their situation, which appears to elude some of the real-life ape language researchers of our modern era, at least in their public statements. On the one hand, many of these ape language researchers have celebrated how humanlike the apes are. On the other hand, these same researchers don't dwell on the fact that the apes, like Red Peter, are captive and doing the work of interacting with humans only because they have no choice.

Intentional Communication

The exhaustive training and testing of Kanzi and Chaser show that when motivated, these animals can be trained to get meaning out of some human speech and react appropriately. The number of different words they can respond to goes well beyond the few predator calls that make up most cross-species understanding in the wild, but it's also limited to at best the understanding of a human toddler. The differences between dog, bonobo, and human child lead to the question of whether Chaser's and Kanzi's understanding of what's said to them differs from how we humans understand talk. To get at that question, we must grapple with what we mean by *understanding*.

When we talk about whether we understand someone, we're usually thinking about understanding someone's language. A friend says that they'll come by at noon, and we understand that they're setting a plan to get together. Language so dominates our lives that it's easy to forget that we can also understand sounds and sights and events in our world without language. If you hear the doorbell, you understand that someone is at the door. You might use language to think about whether it's likely to be your friend arriving at noon or whether it's a package delivery, but you don't need language to associate the doorbell with a visitor. In that sense, the doorbell sound has meaning for you, because it allows you to *predict* that someone is at the door.

Animal communication experts often talk about meaning and prediction in a similar way to this doorbell example. Hearing a monkey's *eagle* call allows many species to predict an eagle in their environment. The *eagle* calls therefore have meaning for those who associate the call with the appearance of an eagle, either via inborn instinct or by learning.

Notice that predictions like these don't have to come from deliberate communication. Betsy the dog becomes wildly excited and runs to the door when her human says, "Walk time!" She behaves in exactly the same way when she hears the jingle of her leash being taken off its hook. Betsy gets the same meaning and makes the same predictions from these two sounds, even though only one of them is intentional communication and the other is just an incidental noise. Similarly, Diana monkeys and their neighbors treat predator calls as important predictors of approaching predators, but they don't necessarily see these calls as deliberate communication signals. Kanzi and Chaser and other animals trained to recognize words also don't need to recognize humans' speech as deliberate communication. In all these cases, what matters to the animal is what is predictive, not what is intentionally communicative.

That's an important difference in understanding between us and these other animals: We know that when someone talks to us, they are deliberately communicating. We can draw additional meaning from recognizing deliberate communication. If a friend says they'll come by at noon, we can not only predict their arrival but also draw many other inferences that stem from our understanding that they are deliberately communicating information to us. We can infer that they want to see us, that they might stay for lunch, that we should maybe put on a clean shirt. Our understanding of intentional communication allows humans to go beyond other animals' simple predictions of what's going to happen next.

We humans even know about intentional communication before we can talk. Researchers have observed that human infants make deliberate communicative vocalizations, even if they haven't yet mastered saying any words. When infants make these sounds, they look carefully at nearby adults to see if the adults are paying attention to them. The infants are showing that they are deliberately engaging and communicating with the adults even before talking kicks in. And when young humans become talkers, their own responses clearly demonstrate their understanding of intentional communication. A parent says, "Nap time!" and the toddler Maya howls "Nooooooo!" This exchange shows that Maya goes beyond predicting what happens next. She also understands that her parent is deliberately communicating a plan, and she quite intentionally communicates her own displeasure.

And feeling more cheerful after her nap, Maya says, "Yes!" when her parent says, "Do you want some apple?" Again, Maya recognizes her parent's talk as deliberate communication, and she joins the conversation. Humans are able to treat vocal signals not just as informative sounds like doorbells; they are able to understand that someone is deliberately trying to tell them something. This extra knowledge yields a level of language understanding that is not available to other species.

Why Animals Don't Talk Back

Little Maya's ability to engage in conversation brings us to the question of what kinds of talking these other species can do after they've been trained to understand some human talk. Dogs and bonobos clearly aren't matching us in the richness of understanding, but they are still doing a recognizable subset of what we can do. There's real continuity between humans and these other animals on the understanding side of communication, meaning that language-trained animals and humans are at different points on a scale of understanding skill.

The talking side is a completely different story. In talking, there's no continuity between humans and other animals in their talking skill. There's only an unbridgeable chasm between us and the rest of the animal kingdom. We're the only ones on the talking train.

Most humans who can hear tend to equate talking with producing speech. As I noted in the introduction, I'll use *talking* to refer to all forms of language production—speech, signing in a sign language, and writing. Researchers in the 1940s who first investigated whether apes could talk to humans also thought of talking as speech, and they attempted to teach an infant chimp named Viki to speak English. The project was an utter failure. Despite years of training in which the researchers manipulated Viki's jaw to form words and required her to vocalize to receive any food, Viki managed to produce only a couple of words.

We now understand several reasons why trying to get chimps to produce speech was going to be impossible. Chimps' vocal tracts—the throat, nose, and mouth components responsible for our speech—are different from ours, and it's currently controversial whether those differences make chimps incapable of producing most human speech sounds. Even if the vocal tracts turn out to be capable of producing speech, chimps' brain circuitry holds them back. Apes don't have brain networks that control spontaneous vocalizations. Their calls, like the Diana and Campbell's monkey alarm calls, are all handled by a different network of brain cells that controls innate, hardwired vocalizations.

We humans also have this hardwired vocalization brain network. It's the one giving us our reflexive "Ahhhh!" cry when we're surprised. I was reminded of this kind of vocalization when I was walking with some of my family under a bridge, and unbeknownst to us, there was someone on the bridge above holding a plastic cup containing the remains of an iced coffee. We were walking and chatting below, and as we emerged from under the bridge the person above released the cup, which fell and hit me in the back of the head. This event was only mildly painful but very surprising, and I immediately let out an involuntary "Ahhhh!" That vocalization instantly got the attention of my family, and they gathered round to see what had happened. These hardwired vocalizations are extremely adaptive for humans and other animals living in social groups, as they alert others of danger or signal them to help.

Of course, humans don't stop with these involuntary vocalizations. In the iced coffee incident, the people who heard my reflexive "Ahhhh!" didn't produce more reflexive calls. Instead, they said many other things, like "What happened?" "Are you OK?" and "Ew, you've got iced coffee in your hair." We humans can go beyond the reflexive vocalizations because we possess two other kinds of brain circuitry for speaking that nonhumans lack. One set of brain circuits allows us to make voluntary vocalizations, effectively allowing us to speak spontaneously whenever we want to. The second brain network makes it possible for us to learn to imitate the speech sounds we hear, so that we can learn the words and sounds of the speech community we're born into. These two brain networks are essential to our abilities to speak with one another, but they're missing from all other primate species.

Without these brain networks for spontaneous speaking and

imitating speech, the chimps and bonobos who can learn to understand some human speech can never speak themselves. In the case of the infant chimp Viki, those thousands of hours of training, pushing her jaws and lips into position and withholding food until sound came out of her mouth, never yielded a speaking ape.

Because primates can make voluntary hand movements, other researchers tried to teach apes to produce some form of talking via their hands. The chimpanzee Nim, who was trained to use manual signs similar to American Sign Language, did learn to produce a number of signs. The signs were all in the service of getting something, and they didn't amount to talking. For example, Nim made the sign *orange* to try to get an orange, the sign *banana* to get a banana, and so on. It was rather like a dog doing a trick to get a treat, but with a different trick for each kind of treat. If making a sign didn't get Nim what he wanted, he made the sign again and again, or he piled on more signs. Some examples include *Banana Nim banana Nim, Eat me Nim,* and *Grape eat Nim eat*. Nim was very smart, and given his deep love of tasty food, he tried whatever signs had been associated with getting that food in the past. If he wasn't trying to get something, Nim didn't sign.

Let's look again at Kanzi, the bonobo who was trained with the keyboard. Careful testing showed that he could understand sentences roughly on par with a two-year-old human child. Throughout his training, Kanzi was relentlessly encouraged to use his keyboard to initiate communication with his human trainers, that is, to talk to them. Sometimes Kanzi would press a key to request something, like an apple, but beyond that, Kanzi and his keyboard said almost nothing. Kanzi might comprehend language roughly like a two-year-old human, but he sure didn't talk like one.

The researchers who trained Kanzi summed up the ape non-talking situation this way: "It is easy to train an ape to say *apple* in order to get an apple but difficult to teach it to use *apple* to describe a food that it is not allowed to eat, a food that it sees someone else eating, a food that it does not like, a food that is found in a particular location, etc." We're seeing the motivation angle again: apes will make hand signs or press keys on a keyboard in order to get rewards, but beyond these clever tricks, they don't use their signs or keyboards for intentional communication. They don't talk about anything. Why?

Here are two theories about what's going on here. One theory is about a social aspect of using language, and the other is about knowing how to use the tools that language gives us. On the surface these theories seem very different, but I think they are closely related, and both connect to the primate researchers' quote about apes using words like *apple* only to get rewards.

One theory, developed by primate researchers, is that apes lack the ability to inform. According to this idea, apes are uninterested, unmotivated, or unable to inform someone of something, except by using their fixed, innate calls or gestures. This lack of informing is likely related to the apes' inability to understand that humans' communications are intentional, because intentional communication is often used to inform.

My husband, Mark Seidenberg, spent time with the chimpanzee Nim during his sign language training, and he wrote scientific papers about Kanzi's and Nim's behaviors. He's told me stories of what Nim was like around his two favorite foods: spaghetti and pancakes with syrup. When Nim saw one of these foods, he hooted wildly and displayed intense excitement. Nim was making food calls, which chimps in the wild make when they spot desirable food. Nim had no experience in the wild, but he didn't need any. Food calls, and all other chimpanzee vocalizations, are innate, and chimps need no experience to make them. Nim's food calls illustrate how informing works in nonhuman primates—innate behaviors like postures, calls, and gestures, hardwired in the brain, are triggered by perceiving food or something else important in their surroundings.

We also saw this behavior in the Diana monkey predator calls—hardwired in the brain, present from birth, triggered by detecting a predator. Even though chimps are far more cognitively advanced than monkeys, informing doesn't extend beyond these hardwired behaviors in these apes. We don't yet know why.

The second theory is about a part of language that the language-trained apes can't master. It was proposed by language researchers, including my husband, Mark. The idea is that the language-trained apes weren't able to understand that signs or keypresses were supposed to refer to something. The apes understood that the sign for *apple* predicted getting an apple, but they didn't understand that the sign was actually a way to talk about the apple. That's why the apes never used the sign or keypress for *apple* to describe an apple that someone else was eating—they had no concept that the sign was referring to anything or informing anyone about anything. The signs and keypresses were nothing but tricks to get a reward.

Here's a demonstration of the human referring ability: Remember Betsy the dog, who could predict she was going for a walk when she heard either the jingle of her leash or her human's speech? Betsy is like a lot of dogs, except in one respect. She doesn't exist. I chose to talk about a dog and give her a name in the process of informing you about the nature of animal predictions and understanding. My naming Betsy and referring to her are examples of what we humans do in the process of intentional communication to inform each other—we use words like *Betsy*, *dog*, and *understanding* to refer to things and ideas. When we inform and refer, our words don't get automatically triggered by some stimulus, as with Diana monkeys' *leopard* calls or Nim's food calls. Our words convey our internal ideas, and the physical signal—speech, sign, writing—refers to these ideas. We can even refer to something that's not in front of us or even something that doesn't exist. We're referring to you, Betsy.

Let's try to put all of these ideas about informing and referring together. Apes appear to be unable to inform others except by their hardwired calls. They also lack an ability to refer to something, which is an essential component of informing. These gaps in the apes' repertoire give us the behavior described by the researchers working with Kanzi, that apes will produce the *apple* sign or keypress to get an apple, but they will never refer to an apple or inform anyone about an apple. That's both of the theories right there—no interest in informing, and no ability to refer to things, which is an essential part of informing. I don't know which came first in this chicken-and-egg situation—lack of informing or lack of referring—but it's clear that apes have neither of these abilities.

Human toddlers, meanwhile, are major fans of informing and referring. If Maya says, "Me eating apple," she's not trying to get something; she's already got the apple, and she's referring to it in the process of informing us what she's doing. She'll do this even if her speech isn't needed, when we can already see her with her apple. Toddlers are so into talking and informing that they are like tiny sports announcers, narrating their day with both the play-by-play, as in *Me eating apple*, and the color commentary, with expressions like *Yummy* and *Big apple*.

This talking difference between human toddlers and the trained apes could not be starker. From infancy onward, humans have enormous desire to talk and to inform one another, and they do so spontaneously, without explicit training or tangible rewards like food. For the chimps and bonobos, thousands of hours of training yielded no interest in conveying information themselves, and no use of their words to refer to anything.

Apes' lack of ability to inform and to use language to refer has another mind-blowing angle—these gaps are all on the talking side of things, not on the comprehension side. Apes can understand when someone is using reference and informing them of something, such as that the apples are in the refrigerator. And when Chaser the dog heard a command like *Find Inky*, she would go to her toy pile and look for the toy octopus referred to by the name *Inky*. Both of these examples show that these animals can understand reference, even to something that's not right in front of them. It seems that it's not information and reference in general that are uniquely human. The part that is ours alone is the creation side of things—using words to refer to something, in order to convey information. Also known as talking.

The Dawn of Talking

Because these non-talking apes are our nearest living relatives, it's natural to ask when in evolutionary time our ancestors acquired the desire to inform and the use of reference to do it. Alas, we don't know. Some major developments in evolutionary history, like toolmaking, can leave evidence in the archaeological record, but talking leaves no remains to dig up. Perhaps prehistoric cave paintings are one consequence of humans' drive to inform and deliberately communicate to one another. The oldest known cave art is more than fifty-one thousand years old, and even older forms of drawing and painting likely existed well before the few specimens that have been found preserved in caves. So perhaps we can trace informing and referring to at least this far back in time.

Evidence of informing via speaking or gesturing must be indirectly inferred from what can be found in archaeological digs. Fossils provide clues to changes in brain size and other anatomical developments in human ancestors that may be related to spoken or gestural forms of talking. There may eventually be DNA analyses that let us date when our human lineage gained those two speaking-related brain networks that apes lack—the networks for vocal learning and for spontaneous vocalization. Preliminary DNA evidence suggests that the vocal-learning network may have developed about half a million years ago, in a prehistoric hominid species that later split into Neanderthals, Denisovans, and our lineage, *Homo sapiens*. This evidence doesn't say exactly when talking began, but it may point to one of the necessary components.

We also don't know why the need to inform and its associated reference abilities came to the fore in our ancestors. Some have suggested that informing for hunting was essential to language evolution. Maybe, but we should steer clear of explanations that point to one single source. And there are also some arguments against the hunting story.

The first is that this claim is pretty male-centric, taking the view that it's primarily the male activities that need sophisticated information sharing. Hunting may indeed require informing, but so might many other activities that ancient females might have been doing—finding fruit, distinguishing poisonous from nonpoisonous plants, weaving mats, tending children, and so on. There is evidence of group childcare in hominids from sites in Africa and Europe as long as seven hundred thousand years ago. These sites show fossilized footprints of a few adults with a larger group of children, and in some cases there is evidence of toolmaking and other group activities. It seems that both male and female ancient hominids might have many motivations for informing.

And second, despite having no spontaneous informing capabilities, chimpanzees also hunt. In fact, they conduct group raids to hunt Diana monkeys, swarming through the trees and grabbing any monkey who hasn't escaped to the slim upper branches that can't support a chimpanzee's weight. Hunting isn't going to be the single motivation that led to informing. There were likely a cluster of prehistoric developments that pushed us to be the talking, informing, referring experts that we now are.

ALL THE CHIMPANZEE AND BONOBO LANGUAGE STUDIES ENDED years ago, and all the apes were sent to other facilities. The lack of new projects and the closing of the labs are clear signs that the research hit a wall. Talking animals will remain the province of children's stories, mythology, and speculative fiction.

Nonetheless, we can still get some insights out of these projects. The lingering lessons for talking and understanding language are the contrasts between what the apes did and didn't do. The studies did show how clever the chimps and bonobos are. The studies also showed that the cross-species eavesdropping we saw with Diana monkeys can extend to other species being trained to understand hundreds of words of human language, including understanding reference and new information conveyed by humans. And they showed us that despite all these smarts and comprehension abilities, these animals do not talk.

The rest of this book is about our ability to talk. Though we don't get our wish for animals to talk to us, we get something more valuable—all the benefits of being a talking species, one that can spontaneously convert internal ideas into external signals for others to perceive. We'll begin to understand these benefits in the next chapter by considering the non-talkers of the human species—infants—and how their early pre-talking abilities begin to shape their world.

CHAPTER 2

Talk Baby to Me

And she watched this daughter grow, grasp at words as if they were bright things, shove everything the world offered into her mouth, as if to taste it all.

LAUREN GROFF, "MAJORETTE," IN *DELICATE EDIBLE BIRDS*

QUICK QUIZ: HOW MANY NOVELS OR STORIES CAN YOU NAME with babies in them? I don't mean a passing mention of a baby; I mean scenes with an actual baby present. Not too many, eh? You might think babies are scarce because they don't talk and therefore can't contribute to dialogue, but that's not the whole story. Dogs also don't talk, but they get at least as many mentions in books as babies do. I think a major reason babies aren't typically characters in literature is that they are perceived as passive. Babies not only don't talk; they don't seem to do much of anything, including move a plot along.

This view of passive babies turns out to be way off base. Yes, babies are helpless in some respects, but in others they engage with the world around them and change it, thereby changing their own opportunities to grow and learn.

If your own experience with babies hasn't made you think they are actively influencing their environment, you're not alone. Even most child-development experts missed the boat on this one for quite some time, including many of the researchers who study how babies learn language. For decades, the story in child language was that infants and young children are like sponges—the ultimate metaphor of passivity—soaking up the language around them until they're able to digest it and give some back themselves. The big scientific debates during this time concerned whether these little sponges contain certain kinds of inborn linguistic knowledge that help them sort through the language they absorb, or whether they can get by with more general innate abilities to learn and organize whatever they're soaking up.

Those debates still linger, but today, we have tossed out the baby-as-sponge idea. We now see that infants are active little agents shaping their environment. I don't know if that's enough of a change to make babies more interesting players in literature, but these new ideas certainly make babies much more scientifically interesting than their former spongelike characterization would suggest.

Before infants can have much effect on the world around them, they need to put in a little time figuring out how their body works. Infants start exploring what their bodies can do from their earliest moments. They try to kick a leg. Get fist into mouth. Make some noise.

Babies' adventures in noisemaking require them to plunge into the enormous complexity of their sound-making apparatus. The human vocal tract stretches from the nose and lips down to the larynx deep in the throat. It's crammed full of muscles, cartilage, ligaments, and mucus, and it takes an astonishingly long time for us to learn to make all the parts work together. When four-year-olds make common mistakes like pronouncing *yellow* as *lello* and saying *puh-sketti* for *spaghetti*, they're showing us that even *years* of practice are not yet enough to get all the moving parts working in sync.

Little one-month-old vocal-tract explorers are a long way from attempting anything as complicated as *puh-sketti*; they can't even say *puh* yet. At this point, all they can manage are a few grunts and squeals, sort of pre-vowel sounds. They do their reps of squeals, leg kicks, waving fists, and so on with little regard to whether anyone else is around or even whether they can hear themselves—deaf and hearing newborn babies do their early vocal-tract exercises at about the same rate. After about a month of this kind of practice, which is also a month of brain and body development, babies are ready to start using their newfound skills to get social.

Foraging for Connection

Parents are keenly aware that smiling starts at about two months of age, but while they're transfixed by those loopy beautiful grins, they may be missing the fact that pre-talking squeals and grunts are getting social too. By three months old, hearing infants can adjust these squeals and grunts to create early expressions of how they're feeling, with different sounds for interest, excitement, and displeasure. These aren't fixed vocal routines like the monkey calls we saw in chapter 1.

Instead, within the first few months of life, human infants are producing flexible pre-talking vocalizations that are adapted to their immediate situation.

When infants try out some of these pre-talking vocalizations in the presence of an adult, the adult will often make some combination of smiles, eye contact, and sounds in response. When that combination of infant vocalizations and adult responses happens, we've got pre-talking social interaction. A whole new game called vocal foraging emerges for both babies and their grown-ups. The opening moves are ones in which both infant and adult are exploring the space of their vocalizations and checking out what happens as a result.

To understand how the vocal foraging game works, we can think about the original meaning of the word *foraging*. Animals, from tiny bees to colossal blue whales, forage for food. They explore their immediate environment to see if food is present and move on to try some other site if the local area doesn't have much to offer. In making their decisions to stay and eat versus go search elsewhere, animals are unconsciously weighing the quality of food in the current area, the probability (possibly learned from past experience) of finding something better elsewhere, and the energy and effort it would take to go check out that new location.

Vocal foraging is not hunting for food, it's hunting for social interaction, for connection with others. Both infants and their adults—parents, grandparents, and other caregivers—appear to try out speech sounds in the space of what their vocal tracts can do—high sounds, low sounds, vowels, gurgles, and so on. The space metaphor shows up for vocalizations because different sounds are made in different locations of the vocal tract—some vowels, like *ooh*, are made at the front of the mouth, for example, and others, like *aah*, are made farther back.

The players in the vocal-foraging game make a series of moves: producing a sound, observing others' reactions, and doing more of the same sound or moving elsewhere in the vocal tract to make a different sound. If an infant seems to be responding favorably to some speech that an adult made, the adult tends to keep to this same area of speech sounds, continuing to produce sounds that are similar in pitch and other qualities. If the infant cries or seems uninterested in some speech sounds, adults forage elsewhere, trying out different sounds that are made in other parts of the vocal tract.

At least initially, adults seem to have a quite broad range of what counts as a positive response from a baby, including obvious positives such as the infant looking at the adult, making sounds, smiling, waving arms and legs. But adults at this point will also count as positive anything that doesn't result in the infant crying. This research didn't exist when I was a young mother, but I definitely remember going to some lengths to keep my rather colicky firstborn from crying. I was aware of at least some of my vocal strategizing here, but what I didn't realize was that my infant daughter was foraging her pre-talking sounds right back at me, helping to train me to produce engaging vocalizations for her.

That's right—the infants aren't just reacting to adult vocalizations—they're foraging themselves, making sounds and looking for adult reactions. Because adults are not attending to their infants at all moments of the day, every infant sound doesn't get a reaction from adults. Studies of mothers of infants suggest that they tend to respond to one-third to one-half of the infants' vocalizations. This level of response is high enough to teach babies that their pre-talking is generating attention from adults. By five months old, infants start tuning their vocalizations to how the adult reacts. If a sound that the infant produces seems to be generating a positive response from a nearby grown-up—smiles, speech, and other forms of engagement—the infant does more of the same. If the adults are unresponsive, the infant forages from a different part of their vocal space, producing different sounds. And if *that* doesn't work, infants try even harder, making a burst of sounds that might generate a reaction.

These little vocal foragers are not passive sponges, waiting for something to happen to them. Instead, they're busy bees seeking out experiences in their world. The infants' own pre-talking and foraging are encouraging others in earshot to engage with the infant and direct speech right at them.

This back-and-forth between adult and infant only intensifies a few months later, when babies go beyond these early vocalizations and begin another form of pre-talking called babbling. For hearing infants producing speech, babbling consists of sequences of consonants and vowels like *bababa*. For deaf infants interacting with adults who sign to them, babbling takes the form of repeated sequences of simple pre-signs that are similar in some ways to the simple syllables of babbled speech.

Babbling babies are continuing their exploration of what the vocal tract and hands can do, and the babbled sequences gradually get more complicated over time. Babbling is more than just practice and exercise, though; it's also social. Little babblers are strategically looking over at their adults' faces as they babble, checking out whether this grown-up is paying attention.

Babblers are in luck on the attention front. Babbling resembles speech or sign language much more than those early squeals and grunts did, and as a result, babbling begins to capture more of adults' attention. Indeed, many grown-ups find themselves imitating the infants' babbling right back at them. The infant's babbling is also cuing adults to talk directly to the infant, using simple words and familiar sounds that the baby has heard before—these are the ideal situations for infant learning, all prompted by the baby's own pre-talking.

And babies have another babbling trick up their sleeves. By about nine months old, when they can sit up and hold things, they look at a nearby toy or one they're holding and babble at it. Pre-talking to an inanimate object sounds pretty antisocial, but it's actually a real attention-getter. Adults find this behavior incredibly cute. "Oh, he's talking to his teddy bear!" they say. And then they start telling the baby about the object, often emphasizing its name in their speech: "That's your BEAR! Your BEAR! That's your TEDDY BEAR!"

By babbling to an object, the little pre-talker has created a perfect moment for learning words—the baby and the adult are both focused on the same object, and the adult is naming it, often repeatedly. Researchers who have studied this behavior with infants and mothers have found that infants' babbling to objects directly changes adult behaviors. Mothers are more likely to name the object when the infant is babbling at it than when they're just waving the thing around or holding it. And changing the adults' behavior in this way has a real benefit for the baby: The more often adults name what the infant is babbling at, the greater the gains in the infant's vocabulary.

Keep That Babytalk Coming

It's not just the fact that babies are encouraging adults to talk to them. As most parents of infants know, babies also change the *way* adults talk. You've likely heard this special way of talking to infants called *babytalk*. Researchers often use the more formal term *child-directed speech*, but I'm going to call it babytalk because I want to be clear that special ways of talking to infants and children are not uniquely spoken; there's babytalk in sign languages too. Whether spoken or signed, babytalk has obvious differences from the way that non-babies speak or sign to each other. Special ways of talking to infants appear in almost every culture and language, including English, Cantonese, Spanish, Russian, American Sign Language, Israeli Sign Language, and K'iche' and Tzeltal, two Mayan languages spoken by indigenous people in Central America.

We might have assumed that babytalk is something adults just naturally know how to do when addressing a baby, but the research on vocal foraging can make us rethink where babytalk comes from. If adults and infants are foraging in the space of sounds or gestures that they make, looking for good reactions from each other, then it's plausible that unbeknownst to the adults, babies are actively shaping the babytalk that's directed at them. We can better understand babies' role in babytalk by looking at what babytalk is and what it might be good for or not good for.

Considering that babytalk appears nearly universally across the globe, it's striking that parents are sometimes told that babytalk is a bad idea. Actually, I take that back. It's not at all surprising that some people scold parents for babytalk, because for almost any decision that parents make these days, someone will likely tell them that they're doing it all wrong.

And babytalk definitely has a bad rap. In a recent survey of English-speaking adults, over half believed that babytalk harmed language development. A typical objection to babytalk is that it provides the wrong kind of language to the child, because the ultimate goal is for the child to speak in a more adult way, not speak in babytalk.

This is a really stupid argument. A child's life is jam-packed with ultimate adult-level goals, yet the best path for the child to reach those goals is not via adult-level experiences from the get-go. Adults and older children are expected to be able to navigate an intersection to cross a street, but we don't start that learning process by plopping an infant down at a corner and reminding them to look both ways before crawling out into the crosswalk. Similarly, someone might want their child to play a musical instrument, but handing them a saxophone at four months old is not the route to raising a prodigy. Learning all kinds of complex skills, including talking, is more safe, effective, and efficient when the learning experiences are adjusted to the child's level of development. Babytalk appears to be one of those age-appropriate experiences that help infant learning.

The obvious next question is what properties of babytalk are beneficial. The nature of babytalk differs across languages, cultures, and individuals, but there are a few central themes that keep popping up. One is a limited vocabulary—a small number of words that are frequently repeated when talking to the infant, such as *bottle*, *doggy*, *Daddy*, *bye-bye*, and so on. Frequent repetition of these words allows many chances for the infant to figure out what a word means and the situations in which it's used.

A second feature of babytalk is exaggerated intonation in child-directed speech and exaggerated movements in child-directed signing. A common belief has been that these hyped-up forms of talking help children learn words better, because the words are often produced more slowly, with exaggerated articulation.

Maybe, but if that's the whole story, then we need to ask why so many people talk to their pets in something that sounds like babytalk. You've undoubtedly heard some pet owner going, "WHOOOZE a gooood BOOYYY?" to some overjoyed, tail-thumping doggo, where the owner's voice is swooping from high to low pitches and the words are stretched out over time, quite unlike the way dog owners talk to each other. What's the purpose of all that vocal exaggeration? It can't be language learning. Even if this dog is truly a very good boy, he's not going to learn the words *who's*, *a*, *good*, and *boy* from his human's exaggerated speech. Since pet owners must know this, again we need to ask what's the point of this aspect of babytalk.

The research about exaggerated intonation and language learning doesn't provide a clear answer. Some studies suggest that the exaggerated articulation of babytalk benefits learning, and others suggest that exaggerations can distort the speech so much that it can interfere with the baby's ability to recognize the words. While repeating words and using a small vocabulary is likely always helpful for an infant, the benefit of slow and exaggerated speech or signing might vary with the particular situation or the amount of exaggeration in articulation.

Even if the exaggerated speech of babytalk doesn't always help babies recognize the words they're hearing, it does have one clear benefit: exaggerated intonation is very good at getting a baby's attention. Human infants, as well as dogs, cats, birds, and horses, are all known to pay more attention to child- and pet-directed speech than to the speech that we use to talk to other adults. And more of baby's attention leads to more social interaction, more responses from adults, and more learning experiences. Babytalk is the natural result of the vocal and gesture foraging in infants and adults, where exaggerated intonation and gestures capture infants' attention and get incorporated into our own routines of talking to our babies. So, yes, babies are training their adults to give them the babytalk they like!

And, yes, that is a good thing. If you're regularly around babies, don't let anyone tell you not to use babytalk. Whether it's spoken or signed, babytalk is an important signal for the infant to pay attention to you, which is enormously beneficial. The exact form of babytalk that a particular adult uses with a particular infant has been guided by their foraging interactions, which will differ across infants, adults, languages, and cultures. It will also naturally change and fade away as a baby gets older and can talk more. That too is the result of foraging.

Pre-talking Changes Listening

We've now seen the first secret of babies' pre-talking—it gets adults talking, in special ways, directed right at the baby, creating the perfect environment for them to learn about language and about the people around them. It changes adults' behavior. This evidence of babies actively shaping their environment vanquishes the old sponge metaphor, but there's an even more amazing secret function of pre-talking.

This one is not about getting more attention and talking directed to the baby; it's tuning up the baby's brain, changing the way it perceives speech. This assist from pre-talking is a crucial step on the path to learning and understanding language.

In the months leading up to their first birthday, babies who can hear are refining their speech-perception skills to focus their attention on the speech patterns around them. They're getting better at identifying common speech sounds from the language or languages that they hear. As a consequence of this laser focus on the speech in their environment, infants are losing the ability to detect patterns in speech sounds from languages that they never hear. Babies begin life as linguistic citizens of the world, ready to learn any set of speech sounds, but as they near their first birthday, they give up being equal-opportunity listeners and start on the path to becoming local speech experts—better at interpreting familiar sounds, worse at unfamiliar ones.

During this period of major changes in speech-perception skills, babies are pre-talking at ever increasing rates. Could this increased pre-talking have a role in changing their analysis of what they're hearing? We can't conclude that there's any direct relationship between pre-talking and changes in perceiving speech just from the fact that they occur at the same time, because so many other changes are also happening then—babies are getting taller, they're sprouting teeth, their hair is growing, and so are their fingernails. To tell if pre-talking specifically affects babies' ability to perceive speech, we need to dig deeper.

The standard story from the baby-as-sponge era was that the pre-talking is just a bystander in the baby's speech perception. The widely accepted story was that babies' increased focus on the speech sounds in their environment was a natural consequence of just hearing these sounds, and it had nothing to do with pre-talking. It's certainly true that babies must hear the sounds of their own language environment to focus on those sounds and not others. We even know that infants who hear a large amount of speech directed to them are better able to perceive important details of their local speech patterns than babies who hear less speech. The amount of language input definitely matters, but there's now good evidence that infants' pre-talking is working its magic here too.

This idea might sound crazy—how could talking, or not even talking but *pre*-talking, change a baby's speech perception? I encourage you to suspend some disbelief at this point; remember that a major theme of this book is that talking changes our brains, our thinking, and our attention in unexpected ways. Pre-talking affecting speech perception is going to be our first example of this theme.

During infants' speech-babbling phase, babies are not simply rattling off syllables to their grown-ups and toys; they are also making tiny vocal-tract movements while they are listening to other people's speech. Babies aren't generating any sounds with these movements. Instead, they are moving their vocal tracts in ways that imitate how to make the speech sounds they're currently hearing.

Babies never completely grow out of this covert imitation of other people's speech, and you, a former baby, undoubtedly continue to do something like this yourself. To get a sense of how this speech imitation works, imagine listening to a familiar song. You might mentally sing along, even if no sound is coming out of your mouth. Even if your lips aren't moving, you're probably still making some small internal movements of your vocal tract, shifting your tongue and larynx to match the pitches and words you're hearing.

You can detect these movements with a little concentration. Put your hand gently around your throat, just under your jawbone. Now imagine singing a high-pitched *eee* sound—don't actually sing it, just imagine it. You may feel your larynx (visible as the Adam's apple in some throats) rise in your throat with the imagined high-pitched *eee*, and then fall again when you imagine singing *eee* at a lower pitch. And if you imagine alternately saying *eee* versus *ahh*, you can probably feel muscles underneath your jaw moving. These muscles are controlling your tongue to create the mouth positions to produce these different vowels, even if you're not actually saying them aloud.

These examples are designed to alert you to the way your vocal tract moves even when you're not speaking aloud. More subtle movements of your vocal tract happen at least some of the time while you're listening to someone else speaking. Scientists think that for adults, these small movements can be helpful in understanding what someone else is saying. In infants, who are far less skilled at perceiving speech, the boost from these vocal-tract movements might be even more important.

Evidence about the role of pre-talking in speech perception comes from some ingenious studies that used a practical and ethical method to prevent babies' pre-talking for very brief periods of time. The goal was to see what would happen to infants' speech perception if the infants couldn't move their tongues. That's a tricky kind of experiment to pull off, because blocking an infant's mouth or throat could lead to screaming babies, annoyed parents, and a failed research project. But the researchers found a way, and here's how they did it.

The researchers studied six-month-old infants who were just starting to babble. These babies were also about ready to begin teething, when they're generally quite happy to gum on a teether—a soft surface that infants find soothing when their teeth are coming in. The researchers purchased two different kinds of teethers that are popular with parents and babies, and each infant in the experiment was randomly selected to have one or the other type of teether. One group of babies got a teether that was flat, touching both the gums and the tip of the infant's tongue. This teether prevented the tongue tip from moving. The other group of infants got a U-shaped teether that went between the gums but didn't touch the tongue or interfere with any tongue movement. It's brilliant—the babies liked both kinds of teethers and were content to have them in their mouths during the experiment, but one teether limited pre-talking tongue movements and the other didn't.

Each infant sat on a parent's lap, teether in mouth, and listened to several minutes of recorded syllables, like *dadadadadadada*. Syllables with the *d* sound were chosen because we make *d* with the tip of the tongue touching the top of the mouth, behind the teeth. If infants did any imitative vocal-tract movements in response to all these *da* syllables, the tips of their tongues would be expected to be sneaking up to the tops of their mouths, where the *d* sound is made. Infants with the U-shaped teether could make these tongue movements, but the infants with the flat teether couldn't move their tongue tips to imitate a *d* sound.

Occasionally during the stream of *da* syllables, the sound changed to a subtly different variant of *da* that's used in Hindi and other languages. This *da* requires the tongue to be ever so slightly more curved than with our English version of *da*. Infants who can detect this tiny change in the speech tend to perk up, look around, maybe wave an arm or generally show more interest in what's going on, because they just heard something new. For infants who can't hear the change, it just sounds like more of the same old *dadada* they've been listening to, and they don't show any extra interest. The researchers measured the infants' reactions to the sound changes and compared the groups with the different kinds of teethers.

It turned out that the type of teether really mattered for speech perception! The infants who got the teether allowing tongue movements were much better at detecting the *da* changes than the ones that had the flat teether keeping their tongues in one position. This difference demonstrates that blocking infants' pre-talking tongue movements leads to less precise speech perception. Pre-talking tongue movements really are boosting infants' perception and learning.

These results fit nicely with what we know about *sensorimotor integration*, brain processes that link the perception of something and producing a related action. Sensorimotor integration occurs in many different areas having nothing to do with speech. One example comes from research about how people recognize and understand emotion in others. Humans produce several subtly different kinds of smiles and smirks, conveying welcome, amusement, dominance, and so on. Our ability to "read" people's faces and recognize what kind of smile someone is making is important for understanding the emotion behind the smile and the nature of the social interaction. We are better at this smile recognition when we also smile ourselves. Similar to the effect of imitating tongue movements in speech perception, the imitative act of smiling in response to another's smile increases the precision of perceiving that smile correctly.

We can extend this point to babies too, in studies comparing toddlers who routinely used a pacifier and those who did not. You probably haven't tried smiling with a pacifier recently, but it's much harder to smile with one in your mouth than without one. Researchers wondered whether toddlers' frequent use of smile-blocking pacifiers would result in poorer perception of smiles compared to toddlers who did not routinely use a pacifier. They found that toddler boys who frequently used a pacifier in the daytime had lower abilities to recognize smiles than those with no or low rates of daytime pacifier use. While more research is needed, this work is also consistent with sensorimotor integration, where actions—smiling, speaking, and pre-talking—tune perceptions.

And speaking of pacifiers, we can ask whether they interfere with tongue movements and other pre-talking. Pacifiers can be soothing and beneficial when a baby is sleeping, but they do limit tongue movement for talking or pre-talking in waking hours. The evidence isn't entirely clear on whether pacifiers have a negative effect on language development, and the studies have tended to be small. Some research suggests that extensive use of pacifiers during waking hours is associated with slower development of articulation skills and lower vocabulary. For this reason, speech pathologists, who provide therapy to children with language delays, tend to recommend limits on pacifier use. Other studies find weak or no evidence for a relationship between pacifiers and language development. The effect of daytime pacifier use on an individual child will likely vary with other conditions, such as the amount of interaction with adults in their environment and whether the child is at risk for hereditary language delays.

And Then They Talk!

Sometime around their first birthdays, pre-talkers become talkers. At first, single words come out, then maybe half a year later, little talkers start to be able to string two words together, and about half a year after that, they can manage phrases with three or more words. The fact that it can take a whole year to get from single words to saying three words in a row is yet another reminder of how very difficult all the parts of talking are and how much practice is needed to get good at it.

Becoming a talker obviously lets kids communicate better. That's huge, but not really news. I want to focus on other important but much less obvious consequences of being a talker. Talking has benefits for speech perception, as we just saw, but also brain development, attention, thinking, learning, and many other aspects of higher cognition that are uniquely human in their complexity. These benefits and more will appear throughout this book. Right now, while we're discussing quite young children, I'd like to look at the features of a child's environment that tend to support or inhibit the child in practicing talking and becoming a talker.

Most parents have probably heard by now that providing rich language input to children is extremely important. That is certainly true; children whose parents frequently speak to them are able to perceive the speech sounds of their language more quickly and tend to develop larger vocabularies. And down the road when it's time to enter kindergarten, these children are typically far more ready for school than children who received less language input in their earlier years. Sadly, studies of school progress show that this gap doesn't shrink in school; it tends to grow larger over the elementary school years. Strong language skills are needed to smooth the path to becoming a reader and to keep up with the demands of schooling.

Many aspects of a child's environment, including poverty, nutrition, pollution, and so on, affect their development and school readiness. We can't point to language input as the only force here, but there is direct evidence that the amount of language input matters, beyond these other important differences in family circumstances. Some (though not all) programs encouraging and helping parents to talk more with their children have found increased amounts of parent talking and improved language abilities in children, compared to parents and children who were not part of the talk-boosting program.

Given this evidence, it's natural for pediatricians and child-language experts to emphasize parent input and overlook the role of children's own talking in their development. But there's a hitch here, because part of this emphasis on increasing input to the child buys into lingering ideas about babies being passive. The advice for more talk is basically a suggestion that parents should pour language on their little sponge-baby and let them soak it up.

We've seen that babies aren't so passive and that their own pre-talking affects their environment and their language development. We definitely should continue talking to babies and children, but we should also ask how much the child's own talking is driving their language development. And if a child's talking is important, then we can ask another important question: whether there is any evidence that the home environment affects children's talking, beyond the amount of language input they receive.

The evidence here is clear: Yes, children's talking matters, and yes, the home environment affects how much children talk. The amount of back-and-forth conversation that adults have with children is at least as important to children's language development as the sheer amount of language input the child hears. Remember the pre-talking vocal foraging that infants engage in, and their checking on adults' attention when they are babbling? These are the earliest forms of infants getting adults to engage in back-and-forth interactions with them. Children thrive on these interactions well after the vocal foraging phase.

Conversation is enormously important for children's development because it allows children to engage socially and to do some of the talking. Conversations encourage young children to think about what they want to say and to make their language convey what they're thinking. These situations allow the child to generate their own ideas to communicate and work out how they want to realize those communication goals via talking.

Of course, some conversations with children are pretty minimal—an adult asks if the child wants some apple, and the kid says OK. There's nothing wrong with this kind of exchange, but other conversations allow children to generate ideas and the words to convey them. The adult might ask what an apple tastes like: Is it sweet? And then what else tastes sweet like an apple? Depending on the age of the child, there might be mostly one-word answers here, but even these are stimulating thought and talk in a way that goes beyond a one-sided flow of words by the parent.

Sharing book-reading time with children is an excellent way to provide high-quality language input to kids and also to encourage their own talking. On the input side of things, books are great ways to introduce vocabulary and concepts that aren't part of everyday life—dragons, ballerinas, rocket ships, gardens, dump trucks, and countless other people, creatures, things, and events that expand a child's relatively narrow world, spark their imagination, and boost their vocabulary.

Reading a book to a child also introduces *book language.* Even books for very young children are written in a way that differs dramatically from the way we speak to children. Kids' books don't just have different vocabulary; they also have different sentence structures and points of view—scene setting in a story, discussion of the distant past or future, and many other ways of using language that don't arise much in a child's day. Exposure to book language boosts young children's vocabulary, school readiness, learning to read, and the child's own talking skills.

And on the kid-talking side, reading a book with a child offers an opportunity to pause to let children talk, answering their questions and prompting them for their thoughts. A parent reading a book to a child could ask what's going on in one of the pictures, for example. A toddler saying "kitty" when the parent points to the cat in a book may seem pretty trivial, but the pathway from seeing a picture or object and producing its name is actually quite complex, stimulating a wide array of brain areas and building language fluency.

Older children may be able to produce a series of words when asked what's going on in a picture, maybe a complete sentence like *He's hiding the apple.* And even more elaborate responses might come from questions like *Why did Sammy hide the apple when Grandma came in?* Here the grown-up's question provides an example of complex language that becomes increasingly important in school, and it can prompt deep thinking by the child. Adults shouldn't worry if their child simply says they don't know how to answer questions like these—it's hard to know what kinds of questions a child might be ready for without foraging around in conversation space to see what works. Adults should just keep the mix of reading and book conversation coming, because both are important for children's development.

Several programs designed to promote reading books to children, including promotions of conversations during book reading, have found that increased adult involvement in reading and conversations can result in real growth in children's abilities to talk and express their ideas. Most of these education programs are aimed at parents who have lower income, education, and resources, but it turns out that even in families where the parents have higher incomes and college and graduate degrees, there are real differences across families in the amount that parents read to young children. Alas, book reading with young children appears to be declining in all family-income groups compared to only a few years ago.

The amount of work involved in raising young children is enormous, and I hesitate to push more advice on top of the onslaught that parents already face. I definitely remember being so tired that I fell asleep in the middle of reading to my kids. On more than one occasion, I was awakened by a little voice saying, "Mommy, keep reading!" So I don't add these recommendations lightly. The huge benefits of book reading and child-adult conversation, and the fact that it's scarce in some households, all push me to urge families to engage in more of it.

The Dangers of Media

It's tempting to think that media for infants and young children—tablets, adults' phones placed in little chubby hands, television shows, YouTube channels for babies and kids—could have very similar benefits to book reading or other kinds of parent engagement. After all, an adult could read a book to the child about whales, or the child could see a video about whales on an iPad. In each case, the child would learn something about whales, right?

Parents who offer more media to young children appear to spend less time reading books with them, suggesting that some parents may indeed be thinking of media as a fine substitute for shared book-reading time. But it isn't a good substitute. Yes, kids might learn something about whales in both cases, but amassing whale facts is not really the primary goal of book reading and conversation with young children. When we think about the importance of conversation and child talking, the book reading and the video are not remotely equivalent.

Homes now can be awash in media for infants and young children in ways that weren't possible only a few decades ago. In 1970, children didn't start watching TV until they were about four years old, but now many infants start watching media at four months of age. Pediatricians and other experts on child development are concerned about increased media use in young children, because watching media pushes out other kinds of kid activities that are important to their development.

Parents may have positive views about media time for infants and young children because companies actively promote this content to them and make misleading claims. Sometimes these companies convey the idea that if parents play this music or game or app or video for their kids, it'll help make their kids smarter. And there can be practical benefits from some media use—when parents need a break or need to get dinner on the table, turning on some kid media seems not only useful for the grown-ups but also potentially beneficial for the little ones.

Beyond any hype that we might encounter about kid media, we adults are extremely ready to believe it's good for young kids because we love media so much ourselves. If we give a toddler a tablet to watch a video, it can be a chance for us to pick up our own phones. Conversing with a three-year-old, even if you just adore him, is not necessarily the most interesting conversation you'll have in your day. It's also much harder work than scrolling on your favorite apps.

It's completely understandable that tired parents need a break, but it's important to be informed about the effects of media on young children. First, the hype around educational or beneficial media, apps, and games is just that—hype with almost no evidence that the media or products are beneficial for children's development. *Sesame Street*, a very carefully designed evidence-based television program for children, is known to have value for certain ages, but almost everything else doesn't appear to have measurable benefits for children's development.

Second, a lot of this kid-media hype is steeped in the old child-as-sponge myth. If babies and young kids were passive little sponges, then sitting and watching some engaging show or phone app might indeed let them soak up something useful. Adults who unconsciously harbor the baby-sponge view might be concerned about the quality of the media that the kids are soaking up, but they might worry much less about the sheer amount of passive consumption time from a heavy media diet.

The trouble with this reasoning is that as we have seen, infants and children aren't sponges. They're innately driven to figure out how to act on the world and forage for connections with other humans. They learn to walk, pick up things, explore them with their hands and mouth, babble to these things. Eventually they run around, build things, and talk to other people. These activities are crucial for the child's development, and they all grind to a halt when kids are sitting in front of media. When kids are looking at screens, they don't talk, and adults don't talk to them. There's no foraging for connections with a video, even if kids are learning to predict when a cartoon rabbit will pop out from behind a bush and sing a song.

The consequences of kid media on talking and cognition are huge. More screen time in preschoolers is linked to lower language skills and less-rich brain development, compared to children who consume less media. These differences can't be explained away by variation in family income. Talking and other generative activities like pretend play—generating ideas and goals and executing them via language and other action—boost learning. Watching media shuts down these drivers of active learning in young children.

For all these reasons and more, the American Academy of Pediatrics and other groups have advised that infants and young children have no exposure to media, or, at most, very, very small amounts. That's quite a big ask these days, as parents have many other demands on their time, and we're awash with the readily available eye candy of children's media. I don't have any easy solutions for busy parents. I do think it's important to remember that a child's own talking and conversations drive enormous social, emotional, and cognitive development, and then make media decisions with this information in mind.

THIS CHAPTER HAS UNCOVERED THE UNDERAPPRECIATED benefits of infants' pre-talking—more adult engagement and language input; magical child learning moments with joint attention, where adult and child are both focused on the same object and the adult is talking about the object and naming it; sensorimotor integration in which the child's own pre-talking tunes their listening abilities. Pre-talking changes both children's brains and the behavior of adults around them.

So far, I've only briefly mentioned that a child's own talking improves their language development and school readiness. We will see more specific benefits to talking practice in children and adults spread throughout this book. Part of the value of talking practice comes from talking being such hard work compared to perceiving language. Every act of talking is a mental workout that exceeds the mental effort of just listening to language or watching media. That hard work is central to the benefits of talking for shaping our minds. Exactly how talking and conversation are difficult, and how we talkers cope with that effort, is the focus of chapter 3.

CHAPTER 3

The Challenge of Talking

Few things hold the perceptions more thoroughly captive than anxiety about what we have got to say.

GEORGE ELIOT, *MIDDLEMARCH*

SHERLOCK HOLMES, FAMOUS FICTIONAL DETECTIVE, BELIEVED that his brother, Mycroft, could also have been a brilliant solver of crimes if he had not been so impossibly lazy. Mycroft was large and slow-moving, and he avoided all unnecessary exertion. He enjoyed sitting by a window and making ingenious deductions from his close observations of the people and activities he saw outside. But rousing himself to leave home and investigate some far-flung crime scene? That was not for him.

Mycroft did venture out to visit his beloved Diogenes Club across the street from his London apartment. The Diogenes Club was in most ways a typical British gentlemen's club, stocked with many of the usual trappings to please a Victorian gentleman—drink, food, servants, comfy chairs in which to puff a cigar and read the newspaper. It deviated from other clubs in one glaring respect. Mycroft and the other club founders established a rule prohibiting an activity that was a mainstay of other gentlemen's clubs. The rule was so strict that members who disobeyed it were expelled. What was this activity that Mycroft Holmes so deeply wanted to avoid?

Talking. The Diogenes Club forbade *talking*. It's not entirely clear why the club founders made this rule, but I think I know why. Mycroft Holmes, deep thinker, exquisite observer, highly averse to expending effort, had made a brilliant deduction: Talking is hard work.

The germ of this truth about talking first struck me as a teenager while hanging out with friends one evening, utterly sleep-deprived following a disastrous camping trip. I noticed that I could understand what my friends were saying, but I was incapable of adding more than a word or two of my own to the conversation. In classic teenager self-absorption, I thought this situation revealed something broken about me, that I had some unique pathology that kept me from speaking when tired.

Much, much later, when I began to study talking, I realized that this was not an unusual flaw in my own brain but something much more general. Under conditions of exhaustion or stress, our ability to talk is far more affected than our ability to comprehend someone else. Why? Because talking is hard work. Challenges like sleep deprivation take a bigger toll on mentally difficult tasks like talking than on the comparatively easy job of understanding others.

This chapter is about the hard work of talking, which is crucial to understand before I can explain how talking tunes our brains. Talking is a mental workout, and all the usual exercise metaphors and clichés apply, especially "No pain, no gain." Indeed, the educational community has a related saying, that appropriately challenging lessons in the classroom are a "desirable difficulty" for students, because only by being challenged do they learn and grow. This perspective applies equally to talking—the hard work of talking is a desirable difficulty that makes us mentally stronger, in so many ways. That's true for pre-talkers, young talkers, and adults of all ages.

Because it may seem surprising that talking is so challenging, some of this chapter explains why talking is both more demanding and slower than comprehending. We'll also see how we talkers cope with this difficulty, unconsciously adopting all sorts of shortcuts to make talking easier and more efficient. The difficulty of talking and our compensating shortcuts work together to create many of the noncommunicative side benefits of talking.

Talking Is a Brain Workout

Talking is not physically demanding in the same way as running a marathon, but our brains are working much harder when we talk than when we're reading or understanding someone else in conversation. We don't notice the effort of talking because what's going on in our brains is mostly hidden from us. We're aware of what we say, and what other people are saying, and occasionally we're aware when something goes wrong and there's a misunderstanding. Most of the time, though, our brains don't reveal what they're doing or how much they're working.

Another reason why you've likely never heard about talking being harder than comprehending is because both researchers and the popular media don't talk much about talking. We regularly encounter comments or articles about many aspects of understanding language—how babies begin to learn words, why children don't read as well as they should, complaints about trying to understand speakers with accents, scientific investigations on how we get tripped up by ambiguity, and expert courtroom testimony on the interpretation of convoluted legal contracts. All of this research and opinion attests to the very real challenges we have in understanding others and the interesting brain processes that make language comprehension possible. But the talking side of things requires even more effort.

Medical professionals do know about talking being harder than understanding. Doctors and communication therapists consistently see the higher burden of talking in their patients, including children, young adults, the elderly, and especially anyone with a brain injury.

Millions of people the world over have been diagnosed with aphasia, a language impairment that results from a stroke or other brain injury, usually to the left hemisphere of the brain. Aphasia can take many forms—difficulty with finding words, pronouncing them, understanding complex sentences, and many other variants. Two common threads run through almost every patient's case. The first is the heartbreak that the patient and family members feel over the loss of communication abilities. The second is that the patient's brain injury almost always affects their ability to talk more than their ability to comprehend other people. Talking, being harder, is more sensitive to disruption than comprehending someone else.

We can also see that comprehension has an edge when infants are just beginning to use language. Babies start to understand common words like *mommy*, *daddy*, and *blanket* as young as six months old, but they don't begin to produce their first words until around their first birthday. By the time babies are getting these first few words out, they probably understand 150 or more words, and as they're saying more and more words, the number of words they understand is growing faster still. Some of this talking lag comes from the need to master the muscle control for speaking or signing, but that's not the whole story. We go through our entire lives understanding more words than we produce ourselves.

Another reason for talking being harder than comprehending is practice. We get better at almost anything if we practice it, but there's a real imbalance in how much we practice talking and how much we practice comprehending. That difference in practice is huge for babies and young children, who don't initially have much to say, but the imbalance persists for our entire lives.

You can see this difference for yourself. Think about the amount of time you spend producing language in a day. You talk to people, and maybe you also talk to your pets. You might send emails, leave voicemails, post on social media, write some memos, fill out a form, maybe even work on a poem, a novel, a blog post, a letter. All of these activities add up to hours each day during which you are talking in some way.

Now think about the comprehension side of things. The talking you do in conversations typically entails some comprehension in return. Then there's the comprehension time that comes from media, including the shows you stream, social media you scroll, or radio and podcasts you listen to. And more comprehension time comes from reading—books, newspapers, magazines, your favorite websites and blogs, more social media, texts, and the pile of emails in your inbox.

When we add all this activity together, it turns out that adults in the United States spend about twice as much time each day comprehending language as producing it, and people in many other countries probably show similar patterns. Practicing comprehending so much more than talking helps to make comprehending easier.

Talking Is Not Built for Speed

The difficulty of talking and the speed of talking have an interesting relationship to each other. When we get better at some skill, we often get faster at it. We can see this in kids' games like speed-solving Rubik's cubes or stacking cups, where the fun is not just to do something correctly but to do it really fast. If speed-stacking plastic cups is a concept that is foreign to you, consider pausing here to check the endnotes for a video showing speed-stacking competitions.

Talking requires us to move muscles to speak, sign, type, or write, and naturally we get more skillful and faster as we practice. However, there's a crucial difference between getting faster with practice in everyday talking and in games like cup stacking. Honing one's cup-stacking skill means practicing the very same actions over and over. Similarly, speakers or singers can practice the very same script or lyrics over and over, and also get really fast. There are some links to fast talking and singing videos in the endnotes, and just as the cup stackers work faster than the eye can see, these folks talk and sing faster than we can comprehend.

Everyday talking is different from all of these examples. The point of regular talking is not to produce a practiced script over and over but instead to say something new almost every time. And because we're saying something new, we haven't completely practiced it before. Sure, we've said the same words in the past, but likely not in this particular order and never in this exact situation. So normal talking doesn't benefit from practice in quite the same way that repetitive skills like cup stacking do.

We can get some more insight about the relationship between speed and practice by considering the form of talking that is likely the slowest for many of us: texting on a smartphone. Reading a text message is usually trivial, even if abbrev slow u down, lol. But replying is much more work. For folks who didn't grow up with texting and don't text regularly, having to tap out a message on a phone is enormously slower than comprehending a text. Younger, more skilled texters who type with two thumbs are about ten words per minute faster than the older crowd who laboriously tap letters one at a time with an index finger.

The fastest texting speeds are reached by those who use the most efficient strategy for current smartphones—not just typing with two thumbs but also ignoring the phone's word prediction suggestions and not pausing to correct typos. Autocorrect software will catch most typos, and if you're texting fast, it's more effective to keep going than to pause to check out what words the phone is suggesting. But even these texting experts reach a plateau where they can't get any faster, and it's not their strategies or thumbs that are holding them back.

Per Ola Kristensson is an engineering professor who helped to develop the efficient texting system called gesture typing. He's studied speed texters, and he's observed that their thumbs don't reach maximum speed when texting; their thumbs can go even faster in other situations. If speed texters' thumbs can go faster, why can't these folks up their texting game even more? Kristensson believes that the fast texters aren't held back by the physical actions of thumb movements. Instead, he suggests that texting speed is "bounded by our creativity."

I think that Kristensson is almost right. I wouldn't quite call it *creativity*, since most of our texting doesn't rise much above the "Meet u at 12" level of style and substance. But he is right that thumb speed isn't the whole story of why expert texters can't get faster. Brain speed is also a limit. Indeed, the brain doesn't just hold back phone-texting speed; it limits the speed of every form of talking.

Brain speed matters because for all kinds of talking, the brain must first make a plan for what we're going to say. A mental plan spanning at least several words is essential for talking fluently. If instead we planned only one word, syllable, or keystroke at a time and then had to stop and plan the next step, we'd never be able to produce a smooth stream of talk.

The brain works rapidly to plan what we'll say, but it has a lot to do. The first component of planning is finding words. Each of us has more than fifty thousand words stored in our long-term memory, and the first demand of talk planning is to find the small subset of them that we need for the idea we want to express at that moment. It's not fully clear how we convert an idea into words, but it appears to begin with concentrating on the idea we want to convey, which results in relevant words in long-term memory becoming suggested candidates for our talking.

Once the talk-planning system has some words to work with, the second step is figuring out what order to put them in. And third, the brain sets up the commands to arrange the physical movements to produce our talk. I'll describe more about these stages of planning and their consequences for our lives later, but the point right now is that each kind of planning takes time, even for our fast brains.

Needing planning time for talking really slows talkers down. You know that feeling when you're conversing with someone and you realize what they're going to say before they say it? There it is—our comprehension is running faster than the other person's talking, to the point that we can guess some of the words that are going to come next. Some call this mind reading or predicting the future. I don't much like these terms because they imply some kind of magic trick, when in fact it's a natural consequence of talking being harder and slower than comprehending. We can mentally finish the talker's sentences because comprehension is cruising along so much faster than talking that we've got some extra time on our hands, and we sometimes use it to predict what comes next.

Podcasts and videos provide a different kind of example of comprehension speeding ahead of talking. Many podcast fans and video watchers use a setting on the app that plays the audio or video faster than in the original recording. This function exists because audiences want it—we can listen faster than the people can speak. Podcast developers have found that the amount of desired speedup depends on the type of podcast, but in general listeners select audio speeds up to about 1.5 times the speaker's original speaking rate. That result corresponds remarkably well to estimates that we can comprehend speech up to 50 percent faster than we can speak.

Measuring Talking Effort

Now that we've seen several different kinds of evidence that talking is harder and slower than comprehending, it's time to dig into the mental effort required to plan our talking. The effort we expend, and the way our brains cope with the effort, will turn out to have a big role in the way talking affects our lives.

It's challenging to measure mental effort because it's hidden in the brain, but researchers have figured out some tricks to get a better sense of what's going on. There's an ingenious technique that doesn't require assessing the difficult-to-measure activity, like the hidden mental work of talk planning. Instead, it asks how much this mental activity interferes with simultaneously doing a second activity that's much easier to measure.

If talking is difficult and we make some research participants do something else while talking, then people should be slow on this second task or make lots of errors doing it, because their brains are being consumed with the demands of talking. And if comprehension is comparatively easier than talking, then people should be somewhat better at simultaneously doing this other activity while understanding another talker.

My favorite example of this method was developed by Amit Almor, who was previously a postdoctoral researcher in my lab. Now he's a professor at the University of South Carolina. Beyond being an award-winning teacher, Amit has always been a very clever experimenter, as this study illustrates.

Amit and his graduate students used the talking-plus-another-task technique to measure the difficulty of both speaking and comprehending someone else. They invited undergraduate students to bring a friend to the lab to have a conversation. In the experiment, the pair of friends sat down across from each other, each in front of a computer. The friends were told that they could talk to each other about whatever they wanted, but one of them was given a second job to do. The friend with the second job would see a moving dot on their computer screen, and they had to use the computer mouse to keep the cursor as close as possible to the dot. Keeping the cursor on the dot was tricky, because the dot sometimes changed speed and direction unpredictably. The friend had to do this dot-tracking while still being involved in the conversation.

Amit's research team used computer software to measure how many screen pixels separated the moving dot and the cursor at any time, which gave them a very precise measure of when the dot-tracking was going well and when it wasn't. They calculated the accuracy of dot-tracking at many different conversation points—when the tracker was speaking versus listening, starting to speak, finishing their turn in the conversation, listening to the start of the other person's speech, and listening when the speaker was winding down.

You can probably predict the first big result: Overall, folks were much worse at dot-tracking while they were speaking than while they were listening. But the experiment also revealed interesting patterns of how difficulty waxes and wanes during a conversation. On the talking side, beginning to talk about something is very hard, resulting in quite poor performance on dot-tracking. Talking gets much easier when the speaker is close to finishing up what they were saying. Why? Because when we first start to say something, talking already requires us to do two things at once. We are doing all the muscle movements to get the first words out, and we're simultaneously doing the mental processes of planning what's going to come out next. Toward the end of whatever the speaker is saying, it's much easier, because there's nothing left to plan, and the speaker just has to say the final words.

Meanwhile, the listener has a relatively easy job early on, but when the speaker is finishing up, difficulty for the listener spikes, and their dot-tracking tanks. Again, why? It's because the listener senses that the speaker is getting ready to finish what they wanted to say and it's soon going to be time to reply. Now suddenly the listener, who's soon to become the one talking, has several jobs to do at once. They must continue to understand what their friend is saying, start planning what they're going to say in reply, and also figure out exactly when it's their own turn to talk. The reply planning starts early, while the other person is still talking, because if listeners waited until the other person was all done, there would be a long, awkward pause while the listener figured out what to say. Indeed, devising the plan for what is about to be said—picking the words, putting them in order, arranging their pronunciation—usually turns out to be the most difficult component of talking, in part because planning is often being done while also listening or speaking.

If talking makes it hard to follow a dot on a screen, imagine what it can do to driving a car. Many states prohibit driving while texting or talking on a cell phone. Most people assume these laws exist because drivers may take their eyes off the road and hands off the wheel while using a cell phone. Those are very good reasons for prohibiting cell phone use while driving, but that's only part of the story. Studies using a driving simulator rather than a dot-tracking task show similar results: Simply speaking aloud impairs driving, and vice versa.

Brain-imaging research also can tell us about the work required to talk and comprehend language. Functional magnetic resonance imaging, or fMRI, is extremely useful for telling us about effort in the brain. You may be familiar with the more generic MRI technique, which uses magnets and radio waves to create a picture of soft tissue inside the body. These images are helpful for diagnosing many conditions, from torn ligaments to cancers. In the fMRI version, the soft tissue of interest is the brain, and the MRI technique is supplemented in ways that tell us something about how hard the brain is working while we use language.

While they are lying in the MRI scanner, research participants are asked to talk or comprehend language, such as describing a picture or listening to a story. Computer software transforms the MRI images to create a picture of the blood flow that brings oxygen and other nutrients to brain cells. When brain cells are working hard, they need high levels of oxygen and other nutrients, and the brain is able to direct this nutrient-rich blood to the specific brain areas that need it. Blood that's carrying oxygen has different magnetic properties than blood without oxygen, and the sensitive magnets of the MRI machine can detect where the oxygen-rich blood is flowing through the brain. With this technique, we can tell which parts of the brain are working harder than others, by detecting which parts are consuming more oxygen.

Hundreds of studies using fMRI have revealed that just about any activity with language—reading, writing, talking, thinking—requires hard work in a vast landscape of brain territory. A second discovery from brain-imaging studies is that even though all ways of using language get large areas of your brain working, talking is a far more demanding brain workout than reading or understanding someone else speaking.

The research we've seen here—dot-tracking, using a driving

simulator, and measuring brain effort in an MRI scanner—all point to talking being both slower and harder than reading or listening to someone else. The greater difficulty of talking sheds light on many other aspects of language use that would otherwise be quite mysterious—why young children understand so much more than they can say, and why brain injuries disrupt talking so much more than comprehending others. It even helps us understand the pattern we saw in chapter 1, that there is some continuity between humans and other animals in our abilities to comprehend language but a complete lack of talking in other species. Talking is just too difficult for nonhumans.

How Talkers Cut Corners and Fight Back

Though you're not aware of it, your brain does not take the difficulty and slowness of talking lying down. It fights back by using shortcuts to reduce effort, get more efficient, and save time. These shortcuts include unconscious choices of words, pronunciations, intonations, word orders, and more. Together, these shortcuts yield *good-enough talking*—quick choices made during talk planning that don't necessarily yield the absolutely clearest way to say something, but are probably adequate to be understood. Good-enough talking usually requires our audience—whoever we're conversing with or whoever is reading what we've written—to do a little extra work to figure out exactly what we mean. Because comprehension is comparatively easy, our audience has extra time and capacity to do that additional work. They probably don't even notice the small extra effort. Good-enough talking goes some way toward rectifying the imbalance in the effort being borne by the talker and their audience.

Probably the most common type of good-enough talking is using fewer words, less carefully articulated. Zadie Smith's novel *On Beauty*, about a family with three grown children, provides a great example. We'll see one of the children, Zora, briefly in chapter 8, but right now, the father is asking his son Levi where she is. Levi's reply is, "Eyeano, swimming?" Smith writes that Levi's *eyeano* answer is a "strange squelch of vowels" that stands in for *I don't know*. It really does look strange on the page, but it's not at all unusual to hear *eyeano* in spoken English. We say this sort of good-enough pronunciation all the time, and people usually understand us. Shortening our words is a great strategy to reduce talking time and help to balance out the relative speed of talking and comprehending.

And this little *eyeano* conversation has more. When Levi continues with "swimming?" he's cutting more corners, using one word instead of "Maybe she went swimming?" In the case of people who know each other well, these sorts of shortcuts can reduce the time and effort of talking and help the conversation move along more efficiently.

An alternative to omitting or compressing words in speech is using gestures to carry some of the meaning. Imagine you're telling someone about a little office birthday party. You say only, "I brought a cake," but you convey much more through a gesture. At the very moment that you're saying *cake*, you hold your hands in front of you in a shape that shows that the cake is medium-size and round, a modest offering appropriate for a low-key birthday event in the break room. This supplementary gesture is a wonderful time-saver—it conveys all sorts of additional information packed into the time it takes to say *cake*. If that information had to come out in speech, talking would drag on longer.

These strategies to shorten our language and off-load some information to gestures work because of another talking strategy: We rely on our audience to tell us if our good-enough talking is in fact good enough for them. As we're going along talking, our audience often sends us signals about whether we're being clear or not—they nod or say "yeah" when things are going well and look confused or say "Wait, what?" when we're not making sense. This helpful feedback from the audience is everywhere; almost 20 percent of the words in our conversations have this function. With so much information about how we're doing, talking-wise, we don't have to aim for absolute maximum clarity. If our audience can't understand *eyeano* or some other shortcut, they'll let us know, and we will adjust.

Good-enough talking also shows up in how we arrange words into sentences. To see good-enough sentence building, we're going to start with an analogy.

Our modern machines and devices, like automobiles and cell phones, are made from hundreds of parts. An auto plant is not a place where a car is manufactured from scratch. Instead, the plant is where cars are put together on an assembly line, made up of components that the auto company has either purchased or made elsewhere. Many manufacturers use a just-in-time strategy for getting components to their assembly lines. They don't premake or buy large quantities of the parts ahead of time but instead schedule the parts to arrive at the assembly line shortly before they're needed.

Companies find that it's a bit of a high-wire act to get everything scheduled properly so that they don't run out of parts and make the assembly line grind to a halt. But when it works properly, just-in-time strategies cut production costs—if the parts are showing up in the factory right before they're needed, there's no need to build and maintain long-term storage facilities for them. Companies brave the stress of just-in-time manufacturing because of the enormous financial benefits.

Talking is clearly not the same thing as automotive manufacturing, but the just-in-time idea captures some key features of the mental assembly line we use to make our talking plans. By using a just-in-time strategy to assemble the words we're about to say, we don't have to store the words in a mental warehouse, and we save a bundle by making the process of assembling sentences more efficient. What we're saving here is not money but a precious mental resource: attention.

We all know that it's difficult to pay attention to several things at once—imagine how overwhelmed we'd feel if two people were simultaneously asking us different questions, a phone was ringing, and it sounded like someone had just broken something in the next room. We would be overloaded because our attention is limited, and we can't spread it across all these events. That example seems obvious, but what is much less well known is that we need attention not just for what's going on around us but also for many brain processes to work properly. As with our attention to the outside world, our internal attention is limited.

We make the best use of our limited attention by using just-in-time planning to convert plans into talk as rapidly as possible. As soon as some words that we need come out of our long-term memory, we devote internal attention to them and start arranging them in a plan for what to say. And as we get going on this word ordering, we start planning pronunciation. When just a bit of the plan is fully ready to go all the way to pronunciation, we start talking while continuing to plan the rest. All of this embodies the just-in-time idea of assembly—we start planning the words we've got and then get them off the assembly line by saying them so that we can turn our attention to planning upcoming words.

When it works right, this just-in-time system that mixes talking and planning is extremely efficient. But there's a supply chain issue, because to start talking right away, we must plan the first part of the sentence we want to say. If instead we started our planning with the words that come at the end of the sentence, we'd have to keep that chunk of plan sitting around, devoting precious attention to it, until we had planned and said everything that came before those words. That mix-up would defeat all the benefits of just-in-time planning.

The brain has a solution. It takes whatever words first emerge from long-term memory and concocts a sentence starting with those early arrivers. You might think that this sounds crazy—if we started talking with any words that happened to be available, wouldn't up jumbled sentences our be, this like? Fortunately, it's not quite so chaotic. When we concentrate on our ideas to retrieve related words from memory, the words that tend to arrive first are the most common ones, words we've comprehended and said over and over again. Because we've had a lot of practice arranging these words into sentences in the past, starting a new sentence with them is relatively easy. And because it's pretty easy to get going with these words, we can move on to planning the rest of the sentence while these easy words are heading off the assembly line. Starting sentences with common, quickly retrieved words is the *easy-first* bias in sentence planning.

This combination of easy-first and just-in-time planning is an incredible boost to the efficiency of talking, especially in the give-and-take of conversations. Imagine two people having a conversation and tell me this: About how long is the time gap between when one person stops talking and the other one starts? If you said that the gap is pretty short or nonexistent, you're on the right track. In both spoken and signed languages, the pause at the switching point between talkers typically lasts less than half a second. In those rare times when the gap drags on longer, it usually feels pretty awkward. A long gap signals that something unusual is going on—maybe the person who's supposed to talk next is embarrassed, didn't understand the question, or is just rude and not really interested in this conversation.

We avoid awkward long conversational gaps thanks to just-in-time planning and easy-first ordering. While we're comprehending what the other talker is saying, we're unconsciously making guesses about when they're going to finish up, and we begin our own just-in-time planning for what we'll say. Easy-first has us beginning our sentence plan with the words initially available, and we start saying those words right about the time the other person stops. This efficient planning removes long gaps between talkers, and the fluid back-and-forth benefits everyone participating in the conversation.

Planning for Talking and for Action

Conversational turn taking is said to be part of the universal infrastructure that allows us to use language. Turn taking appears before talking itself—we can see it in little pre-talkers and their grown-ups taking turns making sounds to each other. It is not, however, uniquely human, because scientists have observed turn-taking behavior in nonhuman primates, who take turns making their hard-wired calls. Turn-taking abilities thus seem a necessary precursor to many species' communication.

This state of affairs, in which something crucial for talking appears in nonhuman species, got me wondering whether easy-first ordering and just-in-time planning could also be found in non-talking behavior. Talking is a form of action, and we unconsciously develop plans for all kinds of actions, not just talking. I wanted to know whether other actions are also guided by just-in-time and easy-first.

To answer this question, I had to go outside my comfort zone and into some unfamiliar research areas. I started reading articles about how we perform actions like picking up a glass and building a shelf. I agreed to co-teach a class about how humans and other animals coordinate joint actions, because that would force me to learn more about how coordination between talkers in a conversation might or might not be like other coordinated actions, such as ballroom dancing or monkeys taking turns grooming each other. Venturing into another research field is no picnic, and at first the papers I read made no more sense than if I was trying to understand auto mechanics or plant genetics. I was having trouble finding any discussion of action planning that bore even the slightest resemblance to what I knew about talk planning.

Eventually, I caught some breaks. I found what looked like obvious easy-first behavior in navigation experiments, where someone traverses a route to multiple destinations in a room or traces a path on a map. Often the set of locations to be visited forms a rough loop, beginning and ending at the same point. In these loops, someone expends the same amount of total energy if they visit all the destinations in a clockwise or counterclockwise order. But neither humans nor pigeons are random in the direction they go from the start position. Instead, they tend to head to the nearest destination first, which is naturally the easiest and fastest one to get to. And in an interesting wrinkle, when researchers put some barriers around the nearest destination, making it harder to get to, both pigeons and people headed in the opposite direction, going first to a farther but now comparatively easier location to reach. I had found it. It was easy-first without language, without turn taking, and even in pigeons.

Why does easy-first show up here? The answer is the same as in talking. Beginning with an easy subcomponent, in this case a nearby destination, allows the human or pigeon to get going quickly while they continue to plan the rest of their actions.

Eventually I connected with colleagues who wanted to collaborate in investigating the relationship between talking and other kinds of actions, including musical performance. Jazz is often played by a group of musicians, but with improvised solo performances embedded within a song. Musical improvisation is complex, and musicians are under real-time pressure to create interesting solos that also relate musically to the song that the group is playing. These pressures are similar in some ways to the time pressures that talkers face in conversations. My colleagues and I hypothesized that using easy-first and just-in-time planning could be an obvious benefit to jazz musicians, who need to plan their solos on the fly. If so, then they should start their solos by playing easier musical phrases while they continue to plan more complicated parts coming later. This idea turned out to be correct: We found that less complex musical phrases came earlier than more complex ones in expert musicians' jazz solos. There it was again—easy-first outside of talking.

These days I can see easy-first in many different actions, including routine ones you can observe in yourself. When you're brushing your teeth, you're doing a sort of navigation task, in the sense that your toothbrush is supposed to visit all your teeth sometime during the brushing. As toothbrushing is an action, you need some (typically unconscious) action plan to order the parts of the mouth you'll brush. I predict you have an easy-first order here too, starting out somewhere easy to brush and later moving on to teeth that are harder to reach. The spots in your mouth that offer the easiest brushing and the most comfortable way to hold the toothbrush are the outer surfaces of the front teeth and the back teeth on the side opposite the hand you hold the toothbrush with. Studies of toothbrushing behaviors (yes, I went looking for them!) show that people really do follow easy-first here too.

It's interesting to think that actions as diverse as brushing teeth, improvising musical performance, and navigating a route on a map are all shaped by easy-first. Finding easy-first in actions also tells us something about talking. Because we see easy-first in other animals, like the navigating pigeons, we can see that its role in talking is not unique to the language system but part of more general action-planning abilities. We don't know its full extent, but easy-first may naturally emerge whenever there's a need for efficient planning for any kind of action. Talking definitely reaps the benefits of this kind of planning.

THUS BEGINS THE STORY OF HOW TALKING WORKS, AND THE benefits of cutting corners to make our talking more efficient and good enough to be understood. The way we plan our talk—how we pick words, order them, assemble their pronunciations—all have profound effects on our lives. We'll encounter these effects throughout the rest of the book. For example, in chapter 7, we'll see how easy-first and good-enough talking shape all the languages of the world. In chapter 8, we'll see how really clever strategies of good-enough talking are judged by language purists as bad language. And in the next chapter, we'll see how the way we pick our words for talking shapes our attention, our motivation, and our mental focus.

PART TWO

Talking Tunes Our Brains

CHAPTER 4

Talking and Mental Focus

The ancients etched the words
for battle and victory onto their shields and then they went out
and fought to the last breath. Words have that kind of power.

EMILY FRAGOS, "THE SADNESS OF CLOTHES"

ABOUT THREE THOUSAND YEARS AGO, SOMEWHERE IN WHAT is now Iran or Afghanistan, a young priest named Zarathustra sat on a riverbank and had a vision. Well, we don't really know whether he was sitting or standing or lying down having this vision some thirty centuries back, but astonishingly, we do have a record of what happened next. The revelation on the riverbank was that the world had been made by a single divine creator, a force of good. Zarathustra's vision changed both his life and the path of history. He abandoned the assortment of local gods he had worshipped and urged his followers to pursue the righteous path of the one divine creator. His teachings spread across many civilizations and formed the basis of the world's first monotheistic religion, Zoroastrianism.

Today, scholars discuss the degree to which Zoroastrianism

influenced later religions, including Judaism, Islam, and Christianity. Those are fascinating debates, but I want to focus on a different consequence of Zarathustra's riverbank vision—his insights about the power of talking.

In Zarathustra's teachings, following the divine creator's righteous path was to have "good thoughts, good words, and good deeds." These were not three separate goals; instead, the thoughts, words, and deeds formed a triangle, in which each side influenced the others. Good thoughts led to good words, good deeds led to good thoughts, good words led to good deeds, and so on. That's what this chapter is about—the interplay among thinking, talking, and doing.

We'll see how talking to ourselves or to others guides our attention, helps us understand our emotions, sets a path for our goals, improves our health. Talking doesn't just let us communicate. It tunes our brains.

The Talk in Our Heads

Zarathustra mentions both good thoughts and good words, but for considering how talking changes our brains, I'd like to group these two together. Obviously the thoughts we have in our heads may be different from what we say to other people, but it's also true that internal thoughts are at least sometimes very like talking. Many people, though not everyone, report that their thoughts include talking to themselves, or what researchers call *self-talk*. Some self-talk is overt speech or signing in a sign language, perceptible by those around us. At other times, self-talk is internal, in our heads.

Child-development experts believe that we don't start our lives being able to have internal self-talk. Instead, we start only with overt talking, and our abilities to narrate our lives when we are children—*Me having apple. Yummy!* and so on—sets us on a path to develop self-talk. Initially, children's self-talk is still overt and perceivable by others, but it's not necessarily designed to communicate with anyone else. It's for the child's own benefit. Over time, for most people, much of this self-talk becomes internal, not perceivable by others. Beyond being more private and socially acceptable, internal self-talk is also more efficient—it runs faster than overt talking because we don't have to fully articulate our speech or sign language in our heads.

Internal self-talk is nonetheless still talking. We can think of it as an internal plan for talking that's not fully finished or executed, and so it doesn't become overt. Sensitive equipment can detect tiny muscle movements or brain activity associated with our internal self-talk, even though the casual observer wouldn't notice that we're talking internally. And sophisticated computer programs are beginning to turn internal self-talk into external computer-generated voices. This is the start of a revolution for stroke patients and others who can still understand language and who have internal talk but can no longer control vocal tract muscles to speak aloud.

Interestingly, people vary in the amount that they report engaging in internal self-talk. Most of us who use a spoken language can hear a "voice in the head" when we're talking to ourselves or reading silently, but some people say that they don't perceive any internal voice. These differences in internal talk are just beginning to be investigated. So far, it's not clear whether those who report no internal self-talk have a form of self-talk that manifests differently, with no "audible" voice, or whether they have only overt self-talk.

It turns out that self-talk, in childhood or adulthood, overt or internal, is a solidly good thing. We're going to see quite a few examples of its value in this chapter. In many cases, it doesn't seem to matter whether the talking is overt or internal. In other situations, the exact form of the talk seems to have an influence, and we'll see why that may be.

Talking and Executive Function

We'll begin with how talking changes our mental focus and planning, what psychologists call *executive function*. We usually think about executives directing the activities of a corporation, but individuals need to develop plans and put them into effect at least as much as companies do. Executive function is our brain's ability to hold on to plans that we want to achieve and work toward achieving them. Our memory and ability to inhibit distractions operate synergistically to keep us focused on a main goal, turning to subgoals as needed and suppressing irrelevant thoughts that could throw us off track.

Researchers initially thought of executive function as part of basic memory and attention systems in the brain, not necessarily related to language. That view is beginning to change, and now we know that talking, both addressing others and self-talk, is a major booster of our executive function. We, unlike our non-talking primate relatives, can use our talk to sharpen our focus, reduce distraction, and get things done.

Children provide a great entrée into the tight relationship between talking and executive function. Kids start their lives with poor executive function and no talking, and we can track how learning to talk has consequences for executive function. For example, the range of words that a child uses at four years old predicts this child's development of executive function skills some months later. These skills aren't just developing in parallel; the child's growing vocabulary instead appears to be affecting the development of executive function.

More evidence for the link between talking and executive function in children comes from studies of kid-adult conversations. We discussed the importance of kids' own talking for their language development in chapter 2, but there's more to the story: Children's conversation experiences are also linked to their executive function. Researchers study children's executive function with child-appropriate games and puzzles, and they link these skills to the frequency of conversation in the child's life. Kids who engage in more conversations with adults also have stronger executive function skills. These kids are better at staying on task and resisting distractions, even for games and puzzles that have no language components to them.

These results don't guarantee that conversation by itself grows executive function; it's difficult to establish an ironclad causal relationship without unethical experiments, like preventing some kids from having conversations and comparing them to kids who are allowed to talk. Nonetheless, there are plenty of reasons why it is plausible that having conversations promotes children's executive function. One has been mentioned before, that talking is a desirable difficulty that pushes children's cognitive development. Children's talking creates an opportunity for them to multitask: to think of what they want to say, then do the work to find the words, assemble them in order, and get them out—all jobs that require internal attention, memory, and other components of executive function. The mental workout of talking has benefits beyond communicating in the moment.

Another reason is that the back-and-forth structure of conversation directly trains executive function, above and beyond a kid's own talking. As we saw in chapter 3, conversational turn taking requires everyone to multitask. The talker is doing all the jobs necessary to convert thoughts into talk, and others in the conversation are simultaneously understanding what the talker is saying, predicting when their own turn to talk will come, and also starting to plan what they're going to say in return. All that multitasking requires executive function to keep those conversational balls in the air. The more time a child spends in conversation, the more they are practicing these executive function skills.

The benefits of talking for executive function don't stop in childhood. They're with us through our whole lives. Next we'll see the advantages of talking to ourselves for executive function. This kind of talking may be internal or overt speech or signing, but the point isn't to communicate with others; it's to help us focus.

Self-Talk and Getting Things Done

The most telling investigations into how talking changes our thinking, attention, and learning tend to have a key feature in common that lets us infer whether talking is directly affecting our cognition. In these studies, research participants are asked to play some game that doesn't really need any talking or language, like finding some nonsense shape amid many others on a computer screen. While the participants are playing the game, experimenters are studying whether their talking influences their performance. Sometimes the researchers explicitly tell the participants in the experiment to engage in overt or internal self-talk, sometimes self-talk just shows up spontaneously when people are playing the game. But in all cases, the researchers are trying to understand how talking to ourselves changes our ability to get things done. Here's a sampling of results:

- People who are searching for products on grocery store shelves perform better when they're saying the name of the product aloud while searching than if they're not talking.
- Five-year-olds who are trying to track an unfamiliar shape as it moves around a screen full of other moving shapes are better at watching their special shape if they have been taught a name for it. Many children spontaneously named the shape aloud as they were watching it move around; some kids this age often have whispered self-talk and internal self-talk too. Though little kids are just getting going on talking and executive function, having a name to say in self-talk helped kids track their shape.
- Adults who are trying to discover how to sort unfamiliar objects into piles, via trial and error with feedback about mistakes, do much better in figuring out what the sorting criterion is if that criterion is an easily nameable property, like the color red, than if the important property is not easily nameable, like gray-green. The sorters aren't getting any

> feedback in words, just thumbs-up or -down on the correctness of their guesses, but a format that facilitates self-talk—something that's easily nameable—leads to better performance.

All of these little games require close scrutiny of some scene to get the job done—to find something, track something, sort something. The games can all be played without any talking or even knowing a language—variations of them are routinely done by macaque monkeys and other species in studies of how nonhuman primates can pay attention and solve problems. But in humans, the activities that in principle need no language are nonetheless completed faster and more accurately with some self-talk.

This might seem like a pretty surprising result. We saw in chapter 3 that when people had to track a moving dot on-screen while also carrying on a conversation, the demands of talking made the dot-tracking worse. Why doesn't talking also screw everything up here? A major reason is that in the conversation study from chapter 3, talking with another person was completely unrelated to the dot-tracking, but in these games, the talk was self-directed and relevant to the task the person had to do. In those situations, self-talk turns out to be beneficial.

This research shows us a really important point: Self-talk, which is not for communication, tunes our brains to be better at perceiving, paying attention, focusing, learning. By self-talking, we do self-brain-tuning. That gives us a huge advantage over the non-talkers of the animal kingdom. Self-talk is an accelerator of human cognition, even for jobs that don't need language.

So if you've got a quick job to do that requires some attention, learning, and concentration, you might help yourself with some self-talk. You could try saying "Keys, keys" when searching for your keys amid the various other stuff lying around. You might say "I need the orange line" when you're reading the map of an unfamiliar subway system. And when the excitement of a new furniture purchase fades with the daunting task of assembling it yourself, try translating those wordless assembly instructions into self-talk: "Panel B needs to fit into the back using these screws, and the edge with the holes needs to face up." Do your self-talk overtly or internally as you like; later we'll see some examples of long-term benefits of self-talk that probably require overt talking, but in these cases of potential short-term benefits, any form of self-talk should be equally useful.

Why does it work? Being a person who studies talking, I think that this is a fascinating question, and I'm disappointed that there isn't more research to answer it. Scientists have carefully done the first step, showing that self-talk has these benefits, but they have done less digging into why. I don't think we should ignore the "why" question, because it can help guide us to other situations where self-talk might be effective, or even where it might be harmful. So here are my suggestions for why self-talk might tune our attention, our executive function, and other aspects of our cognition. All of them seem plausible to me, but they need more research backing up preliminary results.

Consider the situation of using self-talk to aid finding your keys. To produce the word *keys* aloud or even just in your head, you have to activate the concept of your keys strongly to get the word *keys* out of your long-term memory. The act of concentrating on your keys for the purpose of self-talk may also help you conjure an image of how your own particular keys look. Having that visual image could make your search for the keys more effective. It's another form of sensorimotor integration, where internal or overt talking about something affects searching for and detecting what you've talked about.

Now think about how talking helps us with executive function in these kinds of situations: paying attention, avoiding distraction, analyzing what we see. Part of the way talking does this is by making aspects of our world more discrete. The act of using words (internally or overtly) like *hammer* or *red* changes our representations of what we're naming. By calling something a hammer, we are identifying this object as part of a category that is distinct from other objects, like a mallet. Naming something creates a category and has the effect of making individual examples in the category (all hammers) more similar to one another. Mallets and other objects that are not called *hammers* become more different from hammers by virtue of having a different name.

Having names and discrete categories obviously helps communication—you can say "Give me the hammer" and communicate faster than if you had to describe what you wanted. What's less appreciated is that having names and using them in self-talk can help people attend and learn. We saw one previous example, where people could sort multicolored shapes better when the key feature for sorting was nameable for self-talk (red) than when it wasn't easily nameable (a weird gray-green). Similarly, naming a plan may make it more discrete and important and thereby fix it more firmly in mind, aiding follow-through in achieving that plan.

There's an interesting dark side to how talking makes things discrete. Most of the time, categorizing information with our words helps learning, but in the rare situations in which we want to maintain information in a raw, uncategorized state, talking and its categorization powers can be detrimental.

The clearest example of this phenomenon appears in eyewitness accounts of crimes and other events. Say you witnessed a robbery and are being interviewed by the police. When they ask what the suspect looked like, you give various details, including that they had a round face and brown hair. That description is correct as far as it goes, but in real life, there are many face shapes that could be described as round, and brown hair can mean a huge variation in shade, length, and curliness. Amazingly, by naming the face as *round* and the hair *brown*, you've improved your memory for these discrete *round face, brown hair* features, but you've now become *less* able to remember the exact properties of face and hair beyond round and brown. You might even do worse identifying the suspect in a lineup after saying *round face* and *brown hair* than if you had looked at the people in the lineup before you said these words. Talking and its discrete-making side effects can actually interfere with your memory of the information that you don't mention.

This phenomenon is called verbal overshadowing, meaning that what you say overshadows your memory for what you didn't talk about. It's another example of how talking tunes your attention. In order to describe the suspect, you need to grab some words from memory to convey what you saw. You naturally choose to describe what's easily describable—round face, brown hair. By going through the process of talking about these features, you've made your memory of these features stronger but also more discrete, less nuanced. These talked-about descriptions overshadow the nuance of your memory for other unmentioned aspects of how the suspect looked.

Interestingly, verbal overshadowing doesn't even have to be verbal. A graduate student and I investigated whether gesturing a description, using no language in any form, might also have these overshadowing effects. Conveying information via gestures has one of the key components of talking that we discussed in the introduction—it takes an internal idea (what something looks like) and transforms it into a perceptible signal, in this case hand movements. We found that when people described simple scenes using only gestures, they also experienced an overshadowing effect. Some kinds of scenes tended to be described with complex gestures, and these scenes were remembered more poorly, compared to scenes that could be described with simple gestures. This result shows that an important part of the way that talking tunes our perceptions doesn't always need to involve language. Instead, the act of converting an internal idea into a discrete signal is critical to focusing our attention and changing our perceptions. Drawing a picture, another idea-to-signal conversion that doesn't have language, may have very similar effects.

Talking and Managing Emotions

These same kinds of making-discrete consequences of talking crop up in a completely different domain: engaging with our emotions. Imagine encountering a friend who says, "I just got off the phone with my brother, and I'm so upset!" What exactly is your friend feeling at this point? The word *upset* could mean a combination of many things—shocked, saddened, angry, defeated, resentful, and practically any other negative sensation you could think of. You're naturally uncertain about what your friend is going through.

The upset friend may also be unclear about what emotions are at the forefront of their distress. When emotions are swirling inside of us, it can sometimes be hard to figure out exactly what we're feeling. Attempting to name emotions to others or just to ourselves can make the emotions more clear, more discrete. By trying to be more specific, your friend might realize that *upset* in this case was standing in for feeling disappointed and also worried. From this example, we can glimpse some of the rationale behind therapists encouraging their clients to try to name what it is they're feeling. That information doesn't just help the therapist understand what's going on; the naming benefits the person feeling the emotions, by providing some clarity.

All of this means that when you find yourself in a state of emotional upheaval, the act of naming and describing emotions can be beneficial. You could talk to a friend, therapist, or family member, but this emotion-naming exercise can also be done as self-talk.

This point brings up an important misconception about self-talk, that talking to yourself is useless because you're saying something you already know. Quite the opposite. In this emotion case, someone may not fully know what they're feeling, and by trying to name it, they get clarity on it. This talk-induced clarity in turn makes us better able to react to these emotions, a pattern that's supported by endless studies on the benefits of emotion-naming. We'll take a closer look at one study to get an example of how the emotion labeling works, and how it might work for you.

In this study, people who were terrified of spiders volunteered to complete an exercise designed to desensitize them to spiders and reduce their phobia. The exercise involved gradually moving toward a large tarantula in a see-through container, and in the final step, they were asked to reach into the container and touch the spider. No spiders or humans were harmed in this study, and not every participant could muster the nerve to touch the tarantula or even get near its container. During this exercise, researchers measured the level of the volunteers' emotional response, via skin conductance sensors placed on their hands.

The researchers wanted to understand what kinds of self-talk might be most effective in overcoming the volunteers' spider phobia. They randomly assigned the participants to receive one of four different kinds of instructions for engaging in overt self-talk about their experiences. One quarter of the participants were told to accurately name aloud what their emotions had been when they approached the spider and also to accurately describe the spider. Another quarter were told to use self-talk to downplay both their own phobia and the scariness of the spider. A third group was told to talk about some other topic, and the fourth group was told not to say anything.

A week later, the participants came back and again attempted to approach and touch the tarantula. Their self-talk assignments the week before affected their emotional arousal in this repeat visit: Those who had been told to accurately describe the spider and their emotions while previously approaching the spider had lower emotional arousal compared to people in the other three groups. Also, among the volunteers who were told to describe their feelings honestly, the ones who had followed these instructions most carefully and accurately named their emotions (anxious, scared, etc.) had the greatest reduction in their emotional intensity a week later. These results suggest that making a good-faith effort to accurately name what we're feeling helps us cope with negative emotions.

Brain-imaging methods suggest what's happening in the brain when our talking helps manage our emotions. Naming our own emotions in self-talk or to another person changes the amount of neural activity in the limbic system, a brain network that's involved in detection of emotion and danger. Naming negative emotions like fear and anger decreases activity in this danger-detection brain system, helping it step down from red-alert danger mode. De-escalation of the limbic system lets us better cope with our negative emotions. Talking really does tune our brains.

Another effect of talking on cognition has a somewhat different take on talking and emotions. It concerns bilinguals who learned their second language after early childhood; they are fluent in this second language but not to the same degree as their native language. When they are speaking or using self-talk in their second language, bilinguals are somewhat more logical in their decision-making than when they're using their native language. One explanation for this foreign-language effect, as it's called, is that our emotions and social norms are tied up with the language we learned in childhood. By speaking a later-learned language, we may be able to escape being overly influenced by emotions and habits of our culture. While this research fits our theme of talking, emotion regulation, and brain tuning, I confess that I don't quite know what to make of this work. I suspect that additional research will bring clarity to this relatively new discovery about the effect of talking on problem-solving.

Sports Talk

So far we've seen that talking, including internal self-talk, makes us more able to focus, find things, solve puzzles, stay on task, and manage negative emotions. The naming of emotions and objects also partitions our world into discrete categories that otherwise might blur together—dogs versus wolves, hammers versus mallets, sadness versus despair. Categorizing the world in this way is a huge advantage. It supports our learning, our decision-making, and the way we see the world.

We know about these talking benefits because they've been documented in careful studies in the laboratory. That's an essential step, but we also want to know whether talking gives us all these benefits in the messier real world. Can talking to ourselves really help us pay attention in our jobs, pursue our goals, find our missing keys, and navigate tricky real-life social situations? A good place to look for answers is the part of the real world where self-talk has been most widely embraced—the world of sports.

Hellen Obiri is an elite middle- and long-distance runner who decided to take up competitive marathon running in her thirties. In 2023, she chose to tackle the notoriously difficult Boston Marathon, even though she'd entered only one other marathon race before that. The Boston Marathon field that year was packed with more elite women marathoners than in any previous year, and Obiri, who had specialized in shorter race distances throughout her career, was still learning to pace herself for the long haul. Toward the end of the race, when every part of a marathoner's body is in pain, Obiri used self-talk to keep herself going. Her mantra was *I am strong, I am strong.* She leaned on this self-talk all the way to her first-place finish.

Obiri is hardly the only athlete to use self-talk to reach goals. Self-talk is enshrined in the hall of fame of sports advice, as both athletes and their coaches consider it highly effective in increasing athletic performance for both professional athletes and amateurs.

Googling "self-talk sports" yields about 324 million hits of scholarly research, news reports, videos, websites, and blogs extolling the benefits of talking to yourself, for virtually every kind of sport.

The sports psychology world doesn't get too deep into why self-talk works for athletes, but the benefits for athletes are clearly related to the non-sports examples we've already seen. First, self-talk boosts executive function, helping athletes set goals, bring strategies to the fore, and keep pursuing those goals in the face of fatigue or other setbacks. Hellen Obiri's *I am strong* mantra is a perfect illustration.

Second, self-talk regulates negative emotions, which is an essential skill for dealing with high-pressure competitions. The snowboarder Shaun White benefited from this kind of self-talk when he was facing his final run in the 2018 Winter Olympics. At that point in the competition, White was in second place. All his close competitors had finished their runs, and White's own performance would now determine the ranking of the top finishers. If he pulled off an outstanding score on this final run, he'd be a three-time Olympic gold medal winner, but if his run wasn't absolutely stellar, there would be no gold medal for him. An injury had limited his training before the Olympics, sowing doubts about whether he was still the same Shaun White who had dominated past winter games. A gold medal would silence the critics; anything else would let them rip. At the top of the run, White smacked away the anxiety and doubts and pressure and said the words *Who cares?* He pushed off, nailed the run, and won his third gold medal.

Can you harness this kind of self-talk to improve your ability to regulate negative emotions and reach your own athletic and nonathletic goals? I don't think there's much doubt that self-talk is beneficial in these situations, though you may need some exploration, and perhaps the advice of a coach, to settle on self-talk that works well for you. To aid any exploration you might want to try, here are some research results of self-talk in sports, along with some commonsense interpretations based on what we know about talking.

1. Self-talk in sports appears to be equally effective if it's overt or internal, so do what works for you and your sport.

2. A type of self-talk that seems quite broadly useful for many athletes is instructional self-talk, which focuses on immediate next steps. These self-instructions might involve reinforcing your planned strategy in a match, reminders to follow through on your swing, or other specific performance details that may be getting lost under pressure. Instructional self-talk also can be useful for regulating emotions and fending off distractions, such as *Take a deep breath*; *Focus on the court*; *Ignore the crowd*.

3. Self-talk can have a tone that ranges from encouraging to critical. Some athletes find encouraging self-talk like *You can do it!* very helpful, but others think this tone is artificial and cheesy. Critical self-talk like *Ugh, bad shot* can help some athletes focus on what went wrong and do better next time, but others can feel defeated by self-criticism; they do better with supportive self-talk. Rather like the infants foraging for connections with adults we saw in chapter 2, you

may need to forage in the space of self-talk or find a consultant to figure out what kinds of self-talk messages will be helpful to you.

4. And finally, there's a common thread about self-talk both inside and outside of sporting contexts—self-talk is helpful when the going gets rough but not necessarily when the difficulty shoots through the roof. That's likely because talking itself is work, and the extra effort to produce self-talk isn't worth it when you're really overwhelmed. So if you feel you can't engage in self-talk in some situation, don't.

In every kind of self-talk example so far, we can see that its value is real but also modest. No amount of self-talk will enable me to win a marathon, play the violin, or become an astronaut, all of which I'm utterly unprepared to do. Instead, self-talk nudges us toward reaching goals that are within the realm of possibility for our personal situation.

Michael Cunningham's novel *Flesh and Blood* puts the inadequacy of a self-talk attempt on clear display: "Mary wrote on the envelope the phone bill came in: I will not steal. She put the pledge in writing. But she went on stealing. She didn't know why she did it." Mary is in deep water here, unable to control or understand her behavior. A vow scribbled on an envelope is not up to the challenge of rescuing Mary from a transgression she doesn't really want to give up.

Self-talk can't work miracles, but it's also not nothing. Next we'll see a more intensive form of self-talk with more powerful consequences. It might even be capable of helping someone like Mary the shoplifter. The entire goal of this form of self-talk is to give self-talkers insight into real problems and improve their lives in meaningful ways. As you might imagine, it requires more work than jotting a single sentence on an envelope. It is nonetheless doable by almost anyone, with a little guidance, time, and determination on the part of the self-talker.

The Writing Cure

To this point we've seen self-talk's role in pursuing relatively short-term goals: finding your keys, pushing through to finish a marathon. Self-talk can also have much longer-term benefits, but for these outcomes to occur, both the intensity and the form of the self-talk matter. Writing, which is slower, more effortful, and more suited to reflection, seems to be particularly useful for generating longer-term benefits of talking.

That's why mental health therapists often suggest that their clients keep a journal. The therapist is probably never going to see the journal, but they suggest journaling because this written form of self-talk is known to be therapeutic. One example is a gratitude journal, in which the writer notes the people and events for which they're grateful. This kind of self-talk expressing appreciation appears to have small but real benefits to well-being, especially in improving social interactions with others. As with other forms of self-talk, journaling can focus our attention, bringing important people and events more firmly into view. The endnotes have some suggestions for journaling activities, if you're so inclined.

It's interesting to think about the relationship between journaling and an ancient form of reflective talk, prayer. Although prayer is often internal, those who pray might not call it self-talk. They may instead think of the prayer as being directed to a god, ancestors, or some other audience besides themselves. Audience and religious significance aside, there are clear similarities between prayer and certain kinds of self-talk. Both journaling and prayer can include reflecting on troubling situations, naming emotions, expressing gratitude, and reframing situations in ways that can make the one writing or praying feel better. Studies of prayer show it to have benefits for well-being that are similar to the benefits of journaling.

The story about the power of writing for our mental and physical health leads us to James (Jamie) Pennebaker, a researcher who rather by accident became the founder of this field. Jamie was for many years a psychology professor at the University of Texas at Austin, where he studied the relationship between mental and physical health. In particular, he wanted to understand what made psychotherapy work. There are so many varieties of therapy—Freudian analysis, cognitive behavioral therapy, and all sorts of other flavors. Almost all of them have evidence showing that they help patients, but with so much variability in psychotherapy, Jamie wanted to know what aspects of these different therapies were working, and whether we could use that information to make better therapies. He realized that a commonality across therapies is encouraging a patient to name problems that are bothering them. He found that a patient's ability to name a problem was crucial for the patient to make progress on it. We can understand why—just as naming an emotion makes it more discrete and helps us understand and regulate our emotions, naming a problem we're facing makes it more concrete and provides perspective. It might also boost the executive function needed to take steps forward. That's wonderful, Jamie thought, but not everyone has a therapist to encourage them to dig into a problem and name it. Could these same benefits of the patient talking and naming a problem accrue to people who are talking just to themselves?

This question set Jamie and his colleagues on a path of studying the consequences of self-talk. Over several decades, they've investigated a technique called expressive writing, in which people write about an emotionally fraught event in their lives. The goal of the expressive-writing program is to recall the event, dwelling carefully on its details, the emotions that it evoked, and its long-term consequences. The writing exercises continue for fifteen to thirty minutes for several days. Don't let the short amount of time fool you. This isn't just jotting down a few thoughts and feelings. The exercises require repeated careful reflection about emotionally difficult topics.

In studies aiming to figure out what does and doesn't work in expressive writing, Jamie and his colleagues have randomly assigned research participants to either follow an expressive-writing routine or to write about other topics for the same amount of time and same number of days. They've measured all sorts of evidence of physical and mental well-being before versus after the writing exercises. Compared to the groups who write about everyday topics, the people who complete expressive-writing assignments have been found to have fewer doctor's visits, fewer depressive symptoms, better immune function, better grades, and longer-lasting romantic relationships. The expressive writers report greater enjoyment in life, compared with those who are given a different writing assignment. It's tough to keep track of research participants over time, but when researchers have been able to check back with them later, they've found that some of these writing benefits persisted for months or even years.

Even granting all the benefits for attention, focus, and executive function that I've mentioned, it's somewhat surprising that writing should have all these consequences for mental and physical health. It might have turned out that digging up some old trauma and unpacking it on the page could make people feel worse. But instead, writing about difficult subjects or even small annoyances tends to make people feel at least mildly better, and sometimes very much better. The acts of writing down an event and its backstory, naming its emotional baggage, and identifying its consequences together create a narrative that can make the situation more coherent in our minds. Jamie and his colleagues found that it didn't seem to matter whether the writers thought anyone would read their writing or not. It really is a benefit of self-talk.

"I write," said the prolific essayist Joan Didion, "entirely to find out what I'm thinking." All that writing she did definitely gave her insight about the benefits of writing. Writing is thinking, reframing, and working toward solving problems. It is not a magic wand that removes all of life's travails, and it need not be a substitute for psychotherapy. However, it is an evidence-based step toward improved well-being for many people. If you'd like to give expressive writing a try, there are links to resources in the endnotes.

Writing and Long-Term Goals

My colleague Judy Harackiewicz (pronounced *Hara-KAY-vich*) is interested in motivation—what makes one person persist for years to reach a goal while someone else gives up? One area she studies is the perseverance of students who are interested in STEM fields—science, technology, engineering, and math. There is a huge need for college graduates in these fields, but many students who initially plan to major in STEM end up changing majors somewhere along the way. Some of them probably discover different interests, but other students somehow lose motivation to keep pursuing one of these notoriously difficult majors. Judy investigates using the power of writing to increase students' motivations and persistence in STEM programs.

Judy and her collaborators have investigated whether students' motivations in STEM classes could be improved if students wrote about how STEM schoolwork is relevant for their own lives. In one early study, Judy's team investigated this question for ninth graders taking a science class. Kids in nineteen different classrooms across two different high schools were randomly assigned one of two writing exercises. Half the students wrote about how the material they were learning in their science class was relevant to their own lives and goals, and the other half wrote about some of the topics they were learning in the class. The students also completed a little survey about how much they liked science and what they expected their final course grade to be.

The writing assignment had zero effect on the students who already liked science and expected to earn a good grade in the class. For the students who expected to do poorly, however, this one short writing exercise near the beginning of the term made a big difference. Students who wrote about the importance of science to their lives ended up with better grades at the end of the term and expressed more interest in science, compared to the ones who wrote only about science topics.

To investigate whether these writing exercises could have longer-term effects, Judy and her colleagues followed the trajectories of 2,500 college freshmen who were planning to major in a STEM field. These students had enrolled in introductory chemistry, a required course for almost all STEM majors. Again, students were randomly assigned one or another type of writing. Half the students wrote three essays about the value of the course to their personal lives and goals, and the other half wrote their three essays about other chemistry topics.

Given what we know about self-talk in written form, you probably won't be surprised to learn that the students who wrote about the importance of chemistry to their lives reported more engagement with the course at the end of the semester compared to the other group of students. What's more surprising is what happened next. When Judy and her colleagues checked back in with these same students at the end of their junior year in college, they found a lasting effect of those freshman chemistry writing exercises. The students who wrote about the personal value of the chemistry course when they were freshmen had more consistently stuck with their STEM major in the intervening two-plus years than the students who wrote about general chemistry topics. And this effect of writing on multi-year persistence in STEM was more than three times stronger for Black, Hispanic, and other minoritized students in STEM fields than it was for nonminority students.

This is a big deal. The writing exercises in freshman year, which linked the course material to students' own priorities, shaped the paths of many students' majors, careers, and lives. The news here, along with the studies of writing to improve medical and physical health, is that the effects of thoughtful, reflective writing exercises don't just last minutes, hours, or months; they can last for years.

You know the drill about why—self-talk focuses attention, improves executive function, regulates negative emotions, guides people to reflect on their goals, and increases the chance that they'll pursue them. The stronger benefit for minoritized students in Judy's STEM study probably comes from several sources, including that the writing exercise may have improved their sense of belonging and engagement with the class and the major. The assignment to write about the value of the course for a student's life and goals is itself an affirmation that the writer belongs in the course and their STEM major. That affirmation may be more motivating for students who see very few non-white faces in class and who may initially feel less connected to their class and their major.

Of course, it's not always this easy, and other studies using similar writing exercises have found little or no change in the writers' grades or behaviors. I asked Judy about what might be behind this variability. She had several ideas.

Judy, like Jamie Pennebaker of the expressive-writing exercises, believes that the written form of talk is key for generating reflections that have staying power. I think they're right. The effort required to write, its slower pace, and its lasting signal on page or screen are all likely to increase the degree of insight generated by the exercise.

Judy also thinks timing is important: It may be helpful to give these exercises to students when they're in a state of transition, open to change and trying to chart a new path. The students in Judy's first science study were high school freshmen, and the ones in the chemistry study were college freshmen, all students adapting to a new school and a new stage of life. Students are more open to new ways of thinking during such transition periods, including to thinking prompted by a writing exercise.

Motivation to take the writing assignments seriously is also likely an important factor. For student writing, Judy pointed to the importance of making course instructors partners in the research, so that the writing exercises are real, graded class assignments, likely leading to better student reflection and effort during the writing. Studies in which some researcher just emails students in the course and pays them a few dollars to jot down some thoughts probably will get much less engagement and reflection during the writing, leading to minimal effects on later behavior.

Another way to put this last point is that what you put into the exercise affects what you get out of it. Anyone, whether they're a student, a therapy client, or anyone else, won't glean much benefit from half-baked self-talk. If you decide to follow the expressive-writing plan or do another kind of journaling (with or without the guides given in the endnotes), committing to the program and the effort required will be important to benefiting from your writing.

The Dark Side of Self-Talk

Having now seen many examples of self-talk's benefits, it seems that we can give ourselves a prescription for more self-talk, more writing, more reflection. But as with other kinds of prescriptions, it's always important to consider unpleasant side effects. Self-talk does have downsides, potentially quite serious ones. They appear to arise out of the same brain tuning that gives us the benefits of talking.

Rumination is often described as repetitive negative thinking. It gets its name from the digestive process of ruminant animals like cows and goats, who must rechew their food to have it pass through their several stomachs. Rumination in humans can be thought of as the metaphorical rechewing of previous ideas. Human ruminators continually revisit difficult topics, typically with no new insight or resolution. Rumination can, but doesn't always, appear in many forms of mental illness.

Researchers have worked extensively to understand rumination and its consequences. I think there's a way to reconceptualize rumination that may lend additional insight about some of its effects. Instead of considering rumination to be repetitive thinking, let's call it repetitive *talking*. Rumination is typically self-talk but also can occur as overt conversation with another person. Given what we know about talking's effects on our brains, we may gain some insight into how this repetitive talking can intensify the negative situation of someone with mental illness.

Like other self-talk, rumination often begins as an attempt to solve problems in life or gain emotional clarity. This strategy makes sense, as we know that talking can lead to new perspectives and problem-solving. But for reasons that are not well understood, some people fall into extended repetitive self-talk about their negative circumstances, while other folks do not ruminate excessively.

Although we may not know what makes some people more prone to ruminating than others, we do know a great deal about rumination's powerful negative consequences for the talker: It increases insomnia, reduces confidence, increases passivity. Concentration and executive function decline, leading to poorer problem-solving. No matter what mental health challenges someone has, rumination makes the symptoms worse, longer lasting, and more resistant to treatment.

We can apply what we know about the effects of talking to shed light on these consequences of rumination. We've seen that self-talk is often an intensifier. It focuses our attention on what we're saying, it makes what we're talking about more discrete, it tunes our memories to hold on to goals. If these talk-based accelerators are applied to negative, self-directed preoccupations, then we can imagine that the negative self-talk will have similar, but now harmful, accelerating effects. I'm saying "imagine" here because rumination is not traditionally considered to be part of self-talk, and there is to my knowledge no research drawing parallels between the positive effects of talking and the negative effects of rumination.

Thinking about rumination as self-talk might also clarify the value of different rumination therapies. The traditional view of rumination is that it's a consequence of the disordered thinking from mental illness. From that perspective, improvement in the basic problem, mental illness, should naturally lead to improvement in the symptom, rumination. But now the self-talk view: Given the prevalence and intensifying effects of self-talk in every facet of life, not just in mental illness, it would be important to try to treat rumination self-talk directly to reduce its intensifying effects. Although to my knowledge researchers haven't linked rumination to self-talk, several groups have begun to investigate rumination reduction efforts, separate from treating mental illness. There are some encouraging results: Interventions directly aimed at blocking the intense habit of ruminating self-talk have had more success than therapy that treats the mental illness alone.

Similarly, people with obsessive-compulsive disorder (OCD) are described as having compulsions to do certain actions and also "intrusive thoughts." Perhaps these thoughts can instead be described as compulsive negative self-talk. In John Green's young adult novel *Turtles All the Way Down*, the protagonist, Aza, refers to her OCD thoughts as intrusive weeds, taking over her brain. Aza's internal OCD monologue resembles the kinds of instructional self-talk that can benefit athletes, but in her case it amplifies her anxiety and compulsions: "*You should unwrap that Band-Aid and check to see if there is an infection . . . What if your finger is infected? Why not just check? . . . Time to unwrap the Band-Aid.*" I'm not presuming to diagnose anyone from their self-talk here, especially not a fictional character. My point is that clinicians and researchers might gain insight from understanding the general focusing, concretizing, and intensifying effects of self-talk and apply that knowledge to treatment for negative self-talk, including rumination and intrusive thoughts.

One approach to interfering with negative self-talk is mindfulness meditation. Mindfulness encourages practitioners to focus on the current moment and push away other thoughts or self-talk. Meditators may concentrate on the rhythm of their breathing or observe how different parts of their body are feeling—tense, relaxed, stiff, itchy? Meditation research has found many benefits of these techniques, but from the talking perspective, we can see one clear effect: These techniques replace self-talk with something else for the mind to focus on.

There are still many unknowns about talking's downsides—why some but not all people ruminate or develop intrusive thoughts, who does or doesn't benefit from mindfulness meditation or other therapies that reduce self-talk, who embraces versus gives up on a therapy, and even who engages in any kind of self-talk. These questions point to the need for more research in the understudied area of self-talk and its positive and negative consequences.

Hate Talk

One more downside of talking brings us back to the ancient teachings of Zarathustra. We've seen his triangle of the righteous path of good thoughts, good words, and good deeds, but he also warned against another, darker triangle—evil thoughts, evil words, evil deeds, all influencing and accelerating each other. *Hate speech*, the common term for all forms of talking (speaking, signing, writing) that debase or threaten individuals or groups, seems like an obvious example of one side in this evil triangle, influencing and being influenced by hate crimes and other evil deeds.

Hate speech is the topic of many books, articles, opinion pieces, and podcasts. These address the traumatizing effects of hate speech on the individuals or groups it targets, the relationship between hate speech and hate crimes, how hate speech spreads on social media (brief answer: like wildfire), efforts of social media companies to combat it, the legality of hate speech in different countries and situations. I'm not going to add another long discussion to what is already a mountain of research, policy, and practical writings. Instead, I'll offer one opinion: Hate speech researchers and policymakers haven't been thinking about how talking tunes the brain, but they should be.

The current focus in studying hateful talking is on how one person's hate speech affects the audience that perceives it. These perceivers may be influenced by it, may repeat or forward it, may even harm others. That's a horrific situation that's definitely worthy of our attention. But there's also a noncommunicative type of hate speech, and it too deserves our attention. Hateful talk, even if it's being written in a journal and not communicated to anyone else, tunes the brain of the talker.

Producing hate speech amplifies the belief it describes, further radicalizing the talker's views. The studies behind this conclusion investigate whether talking about an opinion can change the talker's belief in that opinion. This research typically asks research participants to state opinions, often in the form of written self-talk in a notebook rather than communicating with someone else. The researchers vary how frequently the participants are asked to write these opinions in their notebooks. The key result is that the more frequently people state an opinion, the more extreme that opinion becomes. Talking, in this case opinionated self-talk, can be a route to self-radicalization.

Once again, the way talking works is behind this effect. Talking about something, in self-talk or in talk to others, makes the topic more discrete, less nuanced. Talking about something makes it memorable and practiced, making it more likely that the talker will mention it again. More mentions breed more familiarity, which in turn causes these ideas to be viewed more positively than unfamiliar ones. Together, this research shows that self-talk about an attitude intensifies it and makes it seem both more normal and more worthwhile. When we consider that consumers of hate speech may then become producers of it themselves, reposting it or producing their own, it's clear that we need to consider these amplifying effects of not just encountering hate speech but also the intensifying effects on those who produce it. For example, research on the beneficial effects of producing reflections in journaling might be extended to investigate the negative effects of producing hate speech.

THE WIDE-RANGING RESEARCH WE'VE SEEN IN THIS CHAPTER shows the effects of talking on our brains. Whether good or bad, these consequences of talking stem from the fact that the process of converting an idea into talk requires concentration and executive function. The words we choose then make our ideas and feelings more discrete—someone had a *round* face, you're feeling *angry*. Writing offers an extended exercise in bringing ideas and emotions to mind and clarifying them.

The next chapter continues the theme of talking tuning our brains, specifically how talking to ourselves or to others boosts our learning. By now you're probably quite ready to believe that talking could help learning. That makes you very different from an important contingent of policymakers: the educational establishment. Talking in many educational settings, from elementary school through college, is something that educators want to shut down rather than ramp up. Even students think that talking will hurt their learning. We'll look at the roots of this disconnect and consider how we might change minds and help students and educators harness the benefits of talking.

CHAPTER 5

Talking to Learn

> The hippocampus is the part of the brain connected to memory, but I can't remember exactly what that means.
>
> TOMMY ORANGE, *THERE THERE*

IMAGINE THIS LITTLE SCIENCE FICTION–Y SCENARIO: YOU'RE offered a free vacation. It's an incredible deal. You pick any destination, fill your time with the activities you'd most like to do, take whoever you'd like to bring along. Wonderful lodging, food, drink, transportation, all free. No scraping together airline loyalty points, no sitting through a presentation on time-share condos, no obligations at all.

You'd want to know what the catch is, because even in science fiction stories, there's a catch. Here it is: The moment that you and your companions return from the trip, everyone's memories of the vacation will be wiped absolutely clean. You'll never, ever, be able to remember anything about it or even that you went on vacation at all. Every memento of the trip will also vanish—every photo, video, text, postcard, souvenir—*pfft*, gone. And your body will bear no witness—suntan, sore muscles, tattoo—absolutely everything reverts to pre-trip settings. You are completely unchanged by the experience. If anyone ever told you that you'd been on this vacation, you'd think they were crazy.

Would you want this vacation that had no effect on you and that you'll never remember? Me neither.

I'm telling you this story because it illustrates the limited value of experiences if we don't retain memories of them later. That point has an important place in a chapter about learning, because shockingly often, we retain little or nothing of what we were supposed to learn from educational experiences. Just as in the vacation example, you probably wouldn't want to attend a fascinating lecture if you later remembered little more than the fact that you'd been there. This idea isn't science fiction—you probably have already attended classes or lectures that you might have enjoyed in the moment, when you felt that you were learning quite a lot, but they're now a complete blur.

This false feeling of learning is not uncommon. It's been documented over and over, yet it's so powerful that it persists even in the minds of professional educators. From kindergarten classrooms to universities, educators and students flock to ineffective course designs and study strategies in the face of clear evidence that they don't yield much learning. I confess that I count myself among the misguided educators. I truly believed that making my lectures clear and engaging was the path to student learning in the courses I taught. But it wasn't anywhere near enough.

Maybe a dim memory that you enjoyed some lecture is sufficient for you to value the experience, but the educational establishment shouldn't be content with this status quo. Indeed, it doesn't have to be this way. Many researchers and educators are investigating alternative educational experiences that can lead to better learning, in which students lay down lasting memories of what they've been taught and can apply this new knowledge and new skills to other areas in their lives.

Talking is a key component of these alternative approaches. We can take everything we know about talking—that it focuses our attention, facilitates our thinking, keeps our goals in our minds, tunes up our executive function, and everything else—and bring all of it to bear in designing different educational programs for ourselves and for younger generations.

In discovering the educational benefits of talking in this chapter, we'll first look at traditional educational practices, where talking is commonly seen more as a detriment to learning than a boon. We'll see the allure of the learning myths that discourage talking, so that both teachers and students resist including more talking in their learning activities. I'll challenge those myths with the research that shows the benefits of talking for learning and ways large and small to employ more talking in the service of better, deeper, more lasting learning.

The Quiet-Classroom Myth

The basic assumption of much education practice is that understanding some information leads to learning it. The idea is that information is pouring into our brains through our eyes and ears, and if we pay attention, don't talk, and understand what the teacher or book is telling us, we'll develop a lasting memory for the information, also known as learning it. This conventional thinking about how learning works in classrooms is so ingrained in practically everyone's views of education that I encourage you to check your own intuitions and consider how they might be challenged by the evidence we'll discuss. This evidence shows that the conventional thinking is hopelessly wrong. Understanding something in the moment of reading or while sitting in a lecture is not sufficient to induce actual learning, where the student can remember and use the information weeks or years later.

There's a research team in the physics department at Harvard University that investigates the seductiveness of the myth that understanding equals learning. They investigate the beliefs of both teachers and students and write papers with titles like "Measuring Actual Learning versus Feeling of Learning." In one study, they compared the standard college physics curriculum (lecture in class, homework outside of class) with a different approach, commonly called active learning. The active-learning idea comes from research showing that to create real, lasting learning, we need to go beyond understanding the material; we must engage with what we're encountering and act on it in some way. The standard actions of doing homework, taking notes, and reviewing the material are traditional examples of engaging with the information, and each has some positive effects on learning. The active-learning movement is aiming for higher levels of engagement and learning than are often afforded by traditional lectures, homework, note-taking, and the like.

Active learning can take many forms. In the active-learning program in the physics study, much of the class lecture time was replaced with having students work in pairs to solve problems and discuss concepts. Having students work together creates the need to talk and to focus deeply on the issues to make progress during the limited class time available. It also creates accountability between the discussion partners. In this study, every student learned some topics via the traditional lecture-plus-homework curriculum and other topics via active learning. These were real physics classes, and all the students had quizzes and exams assessing their learning and affecting their grades in the course.

The exam grades showed that the students learned much more from the active-learning experiences than from the traditional lecture-plus-homework approach. This result lines up with dozens of other studies showing benefits for active learning in college, in high school, and in younger kids. What's new in this study is that the researchers went to great lengths to probe the students' attitudes about the two different learning methods that they'd experienced.

The students generally disliked active learning. Some resisted getting paired up with some stranger to work through problems. Some disliked working on a problem first before being guided to the correct answer; they wanted the standard sequence of being introduced to a concept in a lecture and doing homework later. And above all, students disliked active learning because they were convinced that they learned more in the traditional lecture-plus-homework arrangement. They were shocked to discover that the opposite was true.

Why do students love lectures? The answer has a lot to do with the fact that talking is harder than comprehending. The physics researchers put it this way: "In large part, the effortlessness associated with listening to a well-presented lecture can mislead students (and instructors) into thinking that they are learning a lot." To paraphrase what we've seen elsewhere in this book, talking is hard work, and students enjoy the easier comprehension experience.

Digging a little deeper, we can trace the understanding-equals-learning illusion to the way our brains let us become aware of some mental states but not others. We usually know whether we're comprehending something, and a good lecture leads to a pleasurable feeling of understanding new material. But actual learning is a different animal from in-the-moment understanding of new information, and this difference has consequences for what aspects of learning we can become aware of.

Learning happens gradually via a brain structure called the hippocampus, which does much of its work while we sleep. The new memories that the hippocampus creates remain fragile for weeks or longer, and their fate depends on what happens well after the initial understanding of the information. Memories of newly learned information get firmly rooted in our near-permanent long-term memory if we're reminded of the new concepts and especially if we engage with them in their fragile recently-encountered state. If not, those weak memories will die off, and there will be no lasting learning. We aren't aware of the progress (or lack thereof) from fragile new memory to more permanent learning because the hippocampus, working nights and proceeding slowly, never sends out a signal about its activities. Without any clear information about how memories are forming, our intuitions about learning are instead shaped by the one clear signal that we do have, the belief that we're understanding a lecture or textbook in the moment. The result is a powerful illusion that comprehending in the moment is sufficient for learning.

In contrast to the relatively easy but potentially misleading experience of understanding a lecture, almost every flavor of active learning requires students to do the harder work of talking: discussing with a partner, writing up conclusions from a lesson, explaining the why behind their thinking. As we saw in the previous chapter, writing exercises like these lead to reflection, insight, and perseverance toward goals. They also lead to learning.

Talking is inherently active learning because it's literally a form of action. We convert our thoughts into words and phrases, unconsciously choosing the words we want and the right order to put them in. Making choices is part of being active, and learning activities requiring the learner to make choices are known to be more effective than activities without choices. Talking naturally builds that in.

In educational contexts, the comparative difficulty of talking ends up being a benefit. Educators love to discuss the concept we've encountered before, desirable difficulty, the idea that a challenging exercise has real benefits for lasting learning. They're right about the value of appropriately challenging lessons, but they err by not realizing that talking itself is a desirable difficulty for learning. Students make the same mistake. The physics students we encountered before liked the ease of following an engaging lecture, but that format's absence of desirable difficulty—the talking and active-learning exercises—ultimately led to less learning.

With this information in mind, we'll next look at specific ways in which talking boosts learning. At the same time, we'll consider barriers from the education status quo that limit including more talking and active learning in classrooms.

Talking and Reading

You probably already know that reading education is in something of a crisis in the United States. The 2022 scores for the test that's called "the nation's report card" show that only one-third of US fourth graders are proficient in reading. Children's poor reading is very worrisome because reading is key to almost every other part of kids' education, eventual employment, and well-being. Low literacy is associated with all sorts of bad outcomes—poor health, poverty, higher rates of crime, and others, above and beyond other social factors such as household income.

When kids fall behind in reading, it becomes incredibly difficult to fix the situation later. If we can find techniques that boost reading early on, that would be a big deal.

Using talking to improve reading education sounds crazy initially, because we think of reading as a quiet, solitary act—sitting with a book or e-reader, reading what's on the page, not talking aloud or being talked to. That's a good description of the goal state, where we want kids and adults to end up when they become good readers. But as we've discussed before, it's a mistake to limit kids' early experiences to ones that mimic the eventual goal. Similar to how parents' babytalk is useful while kids are learning to talk, kids' own talking is useful while they're learning to read.

Talking helps reading in part via sensorimotor integration, the beneficial relationships in the brain between perceiving something and producing a version of that same thing. We previously saw sensorimotor integration in chapter 2, where we discussed how infants' pre-talking tongue movements improve their ability to perceive speech sounds. The idea is similar here: Talking accelerates kids' abilities to learn to understand the written word.

Spelling practice (writing words, a component of talking) is our first example of talking benefiting reading. For many of us, spelling tests were a staple of elementary school and beyond. Times have changed, and kids in elementary schools near you are probably getting less spelling practice than you or I once did. There are two ways to think about this change. One is that because autocorrect software is built into our word processors and cell phones, kids no longer need extensive spelling instruction. The alternative perspective doesn't dispute that we may now need less spelling knowledge than in the past. Instead, it suggests that a major benefit of spelling practice in early elementary school is a boost to sensorimotor integration that improves the process of learning to read.

Studies of spelling practice show that it improves reading in everyone from kindergarteners who are at risk for a reading disability to college students who are already good readers. In young kids, it's an exercise that actively engages them in key parts of early reading instruction, emphasizing the important relationship between the letters and sounds of spoken language. Spelling is active learning—kids have to apply their knowledge to spell words. And spelling creates sensorimotor feedback loops between what they're perceiving—the letters and the sounds—and what they're producing themselves.

Another sensorimotor benefit of talking appears when kids are reading aloud. The goal of reading instruction is to get kids to be efficient silent readers, but once again, experiences that deviate from the ultimate goal turn out to be a beneficial step along the way. When kids can read simple sentences, they need daily practice reading aloud to solidify their reading skills. Indeed, they typically need more practice time than the school day can provide. Schools often tell parents to have their child read aloud to them at home, something like twenty minutes a day. Adults who get this message from their child's teacher may suspect the schools are slacking off, but, in fact, reading is a skill with a long learning trajectory, for which extensive practice is really important.

Adults may also assume that the instruction to read aloud rather than silently is to allow the adult to check that the child is reading, to observe how well they're doing, and to help them as necessary. Reading aloud does have these clear benefits, but that's not the only reason kids should read aloud during the early grades. Reading aloud focuses the child's attention and triggers sensorimotor integration. The child must interpret each word in order to read it aloud, creating a looking, talking, and hearing cycle that further binds the reading components together. Reading aloud twenty minutes a day adds up over time to big gains in kids' abilities.

At least initially, reading aloud is hard work and can be pretty frustrating for both child and adult. Beginning readers naturally read slowly, and reading aloud slows them down even further, making the story they're reading start to drag. I clearly remember resistance to reading aloud in my own kids, and I tried to counter it by suggesting relatively easy books and occasionally reading a few pages aloud myself to keep the story chugging along.

Both spelling practice and reading aloud are examples of active-learning exercises that are difficult in the moment, and that differ from ultimate educational goals. It helps to remember that these desirable difficulties have important benefits for long-term reading skill. The PBS show *Reading Rockets* has good suggestions for reading-boosting activities at home, linked in the endnotes.

The final talking-reading example here reflects our growing understanding that reading success relies on strong spoken-language skills in children. Previously, reading teachers and researchers paid relatively little attention to the differences in children's spoken-language skills. Instead, teachers focused on phonics and other reading-specific lessons as the best path to get everyone reading. The phonics instruction is crucial, but there's more to the story.

Children who have poorer spoken-language skills when they enter kindergarten will not only have more trouble absorbing the basic lessons of reading instruction but also fall behind in reading in other ways. Written language, even in books for little kids, uses very different vocabulary and grammar than the spoken language that kids are familiar with. Children not only must learn how to read individual words but also must learn the unfamiliar patterns of "book language" in order to comprehend what they read. Kids with lower spoken-language skills have a harder time making this crucial leap to the new vocabulary, grammar, topics, and storytelling conventions of book language. That struggle throws up additional barriers to becoming a good reader.

This newer research suggests that we can't ignore kids' spoken-language development in reading instruction. Back in chapter 2, we saw that a child's own talking and turn taking in conversations is at least as important as language input from adults in building a child's language skills. Because those language skills cascade into effects on children's reading, we should consider how to use talking to boost language skills for reading instruction and other school readiness in young children.

It's one thing to say that talking is beneficial for language development and quite another to translate that idea into practice for thousands of children who are at risk of falling behind in learning to read. Fortunately, I can point to one large-scale language intervention that does just that. The program was developed by researchers at the University of Oxford and is called NELI, the Nuffield Early

Language Intervention. It's designed for four- and five-year-olds at the start of formal schooling, with the goal of early identification of children with spoken-language delays and special interventions to improve these kids' language. NELI was first implemented in the United Kingdom and has now been used in more than ten thousand elementary schools there. Some schools in the United States are also beginning to use NELI.

The NELI program builds young children's vocabulary and other language skills through many activities, with a heavy emphasis on the child's own talking. Children listen to stories in small groups and are asked to talk about what they've heard, often in back-and-forth conversation with an adult group leader. A popular component of the talking enrichment is a teddy bear hand puppet named Ted. Sometimes the children take on the role of an older friend to Ted, telling him things and helping him learn. These talking opportunities remind me of research showing that a family's firstborn child tends to have superior language development than later-born children, in part because firstborns have greater opportunities to talk to and teach their younger siblings.

Careful studies of the NELI system, comparing children who were randomly assigned to receive versus not receive the NELI training, show that it results in real gains in children's language skills. Children in families with lower household incomes benefit the most, but the benefits accrue across all children. The training benefits boys and girls equally, and it is equally effective for children who learned English in the home and those who are learning English in school. Although more studies are needed, the available evidence already shows that NELI oral language lessons also translate into better reading skills. NELI's success shows that it's possible to provide talking and language enrichment on a large scale, improving thousands of children's language skills and readiness for school.

Talking and Language Learning

Remember studying Spanish, English, or some other language in middle school, high school, or college? You probably remember that it's not that easy. Our years of tuning in to the properties of our first language make it difficult to learn the pronunciations, words, grammar, and everything else of a new language later on. The difficulty is made worse by the fact that second-language learning in the United States and in many other countries is remarkably aligned with the understanding-equals-learning myth. The turn away from talking in language-learning classrooms is quite odd, considering that talking in the new language would seem to be an obvious component for mastery. Part of the reason for this focus on comprehension is practical: It's relatively easy to have a group of students comprehending the teacher or a video all at once in a classroom, but having each student talking would be quite chaotic.

Another reason behind the comprehension bias in second-language instruction is the theory of comprehensible input developed by Stephen Krashen, an influential education researcher. The comprehensible-input idea is that comprehension equals learning on steroids. Applied to second-language learning, it's the claim that students will learn most deeply when they can readily comprehend the language input they're getting, without the stress of talking in the new language before they feel ready.

You see where this is going, *oui*? The comprehensible-input idea is anti-talking. It emphasizes making the situation easy for the students. If instead we lean into the desirable difficulty of talking and its ability to focus attention and engage active learning, we get a very different approach, where talking is a route to deeper learning of a new language.

A graduate student and I directly investigated this idea. We weren't interested in the obvious point that practicing talking helps folks become better talkers in the new language (though practically speaking, that's nothing to sneeze at). Instead, we wanted to test a much more controversial hypothesis that's the opposite of the comprehensible-input idea. We suggested that talking practice in the new language would help in comprehending it, and that talking would help comprehension even more than comprehension practice itself.

We tested this idea by having college students learn a small made-up language. We invented a new language so that all the learners would be equally unfamiliar with it, no matter what other languages they'd been exposed to. The little language had words and sentences to describe what was going on in cartoon pictures and videos of aliens moving around on their planets.

We randomly assigned half the students to learn the little made-up language by doing comprehension exercises and the other half to do speaking exercises in the language. Everyone got feedback on their performance as they were learning.

We then gave everyone comprehension tests. The group with the speaking exercises, who had never once had a comprehension assignment while learning this language, did better on all the comprehension tests than the learners who'd practiced comprehension. The talking-based learners not only were more accurate, they also comprehended speech spoken in the made-up language much faster than the group that had practiced comprehending.

Getting faster understanding of a new language is a big deal. If you've ever had experience trying out your foreign-language-classroom skills in the real world, you know the feeling that the unfamiliar language is just zooming by when people talk to you, and you just cannot keep up. Getting up to speed is a huge hurdle in navigating real-world conversations and putting your new language skills to use. Talking practice appears to help.

Other studies of the benefits of talking versus comprehension exercises in language learning have found that talking exercises are better than comprehension exercises for mastering the grammar of a new language. The talkers in learning studies also build vocabulary faster and comprehend newly learned words better, compared to students who are assigned to do comprehension exercises. This is bad news for the comprehensible-input idea of language learning. Instead, this research points to the importance of sensorimotor integration, where producing helps perceiving. Indeed, brain imaging studies show that learning new words and other aspects of language depends on communication between the talking and perceiving parts of the brain.

Instructors in second-language classrooms should consider this evidence and look for ways to increase talking in language instruction, at all grade levels. Educational software for language learning should be able to smooth this transition to more talking activities in class and in homework assignments. However, a major hurdle may be instructors' continued commitment to the popular comprehensible-input curricula. It remains to be seen how findings favoring the role of talking in learning will affect teaching practice.

Rethinking How We Teach K–12 Students

Primary and secondary schools have several "traditional" talking exercises in the form of writing assignments and presentations for an audience. Given everything we've seen about the benefits of writing for reflecting, thinking, and organizing those thoughts, it's clear that writing assignments are tremendously useful for children's cognitive development. The act of selecting which information to include or omit, the order it needs to go in, how it should be introduced—all of these subcomponents create focus and decision-making that accelerate learning. Writing and presentation assignments also encourage kids, who are naturally quite egocentric, to think about the information their audience needs to understand what they're saying. That's an incredibly important skill for interacting with others.

These traditional school assignments are some of the original active-learning exercises we have, full of desirable difficulty. We don't yet know how writing assignments will change as teachers grapple with the advent of artificial intelligence programs such as ChatGPT, which can write a student essay or do other homework reasonably well. Given the possibility of artificial intelligence undermining the desirable difficulty of these active-learning assignments, it's now especially important to ask how much active learning and talking are going on in schools.

The answer is not much. Beyond writing assignments, presentations, and activities in drama, speech, and debate clubs, student talking is currently extremely rare in US primary and secondary school classrooms. Most teacher how-tos on student talking offer tips on how to get kids to be quiet in class, not advice about how to get them talking more. It's true that kids' off-topic talking interferes with learning in class, but that doesn't mean that learning-relevant talking must also be banned.

In contrast to abundant studies on active learning in college classrooms, there is very little research about fostering learning-relevant talking for younger students. A few attempts to teach children to think carefully and discuss topics in class have shown benefits lasting for several years after the program. But there are also studies finding only minimal benefits of encouraging discussion. All of these studies are quite small, reflecting the difficulty of conducting large-scale research in actual K–12 classrooms.

Individual schools and classrooms vary in the amount of talking that's allowed. At the extreme end, I've heard of an elementary school with a policy of preventing almost all talking. The school is in a neighborhood where family incomes are low and almost all the students are Black. Two reading researchers who toured the school told me that the school administration was quite proud of their no-talking rules. The school had a silent-hallway policy, a common feature in many US schools. But this school also had a silent-lunch policy—kids had to eat their lunches in silence. And there was a silent-classroom policy, where talking was allowed only in very select circumstances, when the teacher asked a student a question. And there was even silent recess at least some of the time.

I've tried to discover whether this extreme anti-talking stance exists elsewhere, and whether no-talking policies are more common in schools serving non-white children from lower-income families than they are in more affluent, white neighborhoods. Conversely, I'm also interested in whether strongly pro-talking and active-learning policies are more common in well-funded schools serving primarily white students from higher-income families. Unfortunately, it's impossible to find out answers to trends in either pro-talking or anti-talking programs, because policies like these are made at the classroom or school level and are not tracked anywhere. What is well documented is a pattern of school discipline in which non-white students are punished at greater rates than white students. This attitude might be related to a lower tolerance of talking in schools with non-white students.

The irony of banning talking in schools serving poor non-white students is that talking and active-learning activities are known to help close achievement gaps between more and less economically privileged students. Remember that the NELI language-enrichment program, which includes abundant talking, was shown to be most beneficial for children from poorer households. And active-learning activities foster a sense of belonging in the classroom for minoritized students, which in turn improves motivation to do well in school.

Hacks to Increase Talking and Learning

I think that there's room for significant increases in talking and active learning at all levels of education, but radical change does not appear to be on the immediate horizon. In the absence of that institutional overhaul, students, teachers, and caregivers can all incorporate more talking into educational activities inside and outside of the classroom. I'll offer a variety of strategies, all backed by clear evidence for their effectiveness. Most of these can be applied in some way, from elementary school to college, and they are applicable to many different topics. When they're applied in the classroom, they should be structured so that every student does the activity, avoiding situations where only a few students talk and others just look on.

Asking questions

Little kids are famous for asking questions: *Why is the sky blue? What happens to the water when it goes down the drain?* This natural curiosity can be harnessed for learning, and students of any age can be encouraged to ask questions. Forming questions focuses students' attention on gaps in their knowledge and generates more curiosity about what they don't know, prompting them to look for potential answers.

Questions can be incorporated into classroom activities with little extra infrastructure needed. For example, a language arts exercise has young readers imagine they're having a conversation with the author of a book they're reading, and the children's task is to come up with some questions for the author. Studies of this exercise show that it improves children's engagement and reading comprehension, compared with children who didn't do the question exercise.

Generating explanations

The flip side of question asking is producing answers, including explanations. When children and adults generate explanations, they learn more than when they are told the right answer from the start. It's a desirable difficulty that's at the heart of active-learning exercises in which students must work through a problem with a partner. Teachers can prompt students to suggest explanations about practically any topic, at practically any educational level: *Why do bears hibernate? Why didn't the Little Red Hen want to share her bread with the other animals? Why did the United States choose to send troops to Europe during World War I?*

In the spirit of the enterprise here, let me ask you for an explanation: Why do explanations lead to more learning? If you think about what you already know about talking and learning, you probably can identify some of the reasons why the act of generating an explanation has these beneficial effects.

One of the forces behind the power of explanation is the power of talking. We remember things we say ourselves better than what other people say to us. That memory boost for our own words follows naturally from the attention and effort necessary to plan and execute our talk. Another force behind explanation is the power of recalling information. To generate an explanation, we must remember what we've learned. The act of recalling information from memory is active learning and a strengthener of memories.

But explanation can do even more. Generating an explanation improves learning even when the answer is sitting right in front of you. In one study, researchers gave college students worked-out examples of fraction division problems, including a diagram, a list of all the steps to solve the problem, and the final answer. Half of the students were randomly chosen to generate explanations of how these problems worked, using the helpful information given. The other students did different activities with the problems, again using the same helpful information. The students who gave explanations gained more knowledge than the other group, even though everyone had the same information about how to solve these problems. The act of explaining, as with other examples of talking we've seen, focuses students' attention on key components of the problem or concept, resulting in better conceptual knowledge.

Explanation also is valuable in the more common situation in which students have no correct answer sitting in front of them. If a teacher asks why bears hibernate, the need to offer an explanation focuses the students' attention on potential evidence that could be relevant. That in turn may require recall of information from long-term memory—hmm, bears need food and oxygen. And then students must evaluate the evidence: Maybe hibernation is more about food than oxygen?

Explanation also encourages more abstract thinking and more comparison, as part of weighing potential evidence. Kids who offer an explanation for the children's story of the Little Red Hen, whose neighbors didn't want to help her bake bread but want to eat the finished product, will likely realize that cooperation is important for all social groups, not just for the Little Red Hen's community. They get to this insight in part by comparing and talking about the actions of the diligent bread-baking hen and her uncooperative neighbors.

These examples show that explanations tie together many components that lead to lasting learning. The need to explain generates a flurry of mental activity—searching for evidence, recalling information from memory, making comparisons, evaluating evidence, and planning what to say. Each of these components is inherently active learning, and all of them create a route to laying down new memories and consolidating knowledge.

A potentially tricky implementation issue is allowing the students time to develop their explanations. Teachers and other adults may want to rush in with the right answer, cutting off discussion and explanation. Students do need to know what's correct, but it's also important to remember that the process of developing the explanation is itself a valuable learning engine and shouldn't be truncated in favor of a quick answer.

Adding non-language "talking"

Neither drawing nor gesturing is part of our language system, but they share important features of talking. Both involve the transformation of an internal idea into a perceptible signal, such as pointing, sketching, or imitating some action. Given the importance of the idea-to-signal component in talking, it's not surprising that gesturing and drawing can also boost learning. The act of transforming an idea into a drawing or gesture can focus attention on key components of what is to be learned—a math problem, a scientific concept, or any other information that can be conveyed in these modalities. Drawings, graphs, and gestures are natural additions to all the other talking-as-learning strategies that have been mentioned here.

Talking for self-testing

Finally, there's the time-honored practice of students testing themselves on the class material. The test might be an instructor-given practice quiz, or the student might generate their own test-like exercises, including flash cards, explaining the material to a study partner, or attempting to answer questions at the end of a textbook chapter.

Whether a test is imposed or self-generated, the act of digging into memory for the answers to questions has two important effects on anyone's learning. First, unlike our hippocampus, which is silent about whether learning is happening in our brains, the results of a self-test give students information about whether they know the material well. Second, tests are learning events in themselves. Because of the way that our long-term memory works, recalling information to answer test questions makes the memory stronger. Remember that in talking, words that we pull from memory for talking often get advantages over unused words, coming out faster when we need them next. The same process applies to any kind of recall, including remembering a chemical formula, dates, and complex interpretations of political movements. Tests can get a bad rap among students, but the knowledge that tests are actually a driver of learning can put them in a different light.

In my college course called Language, Mind, and Brain, I discussed all these talking-as-learning phenomena and the value of self-testing. To get students on the path of testing themselves, I developed optional online quizzes that the students could use to study for exams. Most students liked the little quizzes and wanted more of them, but they had less success self-testing on their own. Many reported that outside of making some flash cards, they were unclear about how best to test themselves. My suggestions for them were variants of all the strategies I've just mentioned, adapted to a self-testing situation: Ask yourself questions, try to give explanations, try to draw a graph of some results, tell a study partner about the key research you've been learning about in the class, and so on. This brings us to a little insight here: All of these examples of bringing more talking into learning—generating questions, explanations, and so on—are informal kinds of self-testing. Or put another way, the benefits of testing that researchers study—that recalling improves memory, focuses students on what they do and don't know—are natural results of how talking works.

Unfortunately, the self-testing suggestions I made to my students were a pretty hard sell. The students in my class weren't lazy, but they had been steeped in the understanding-equals-learning myth, and their strong inclination was to study by doing more comprehending—reading over their notes, reviewing the textbook, and so on. They were like the physics students in the Harvard study I mentioned earlier—completely devoted to the idea that we learn by comprehending. The act of self-testing seemed difficult, unpleasant (tests!), and overly time-consuming. The researchers who studied those physics students found that some targeted education on the benefits of active learning did improve their attitude about it somewhat. That's an encouraging start, but it will take some time to develop a different learning culture that has moved away from the understanding-equals-learning myth across all levels of education.

Going Large with Talking in Classrooms

The talk-increasing hacks we've seen so far all involve boosting talking via relatively minor tweaks to standard educational practices. A major barrier to their broader use in or out of the classroom appears to be attitudes favoring the lower-effort understanding-equals-learning approach and resisting the desirable difficulty of talking, recalling, and acting to learn. Next, we'll see some programs that move away from business as usual and incorporate some of the active-talking strategies as core parts of their curricula.

An obvious and long-standing example is the philosophy major on college campuses. The educational mission of philosophy courses is to teach students to think through difficult issues and express their thoughts. These classes are structured with ample opportunities for students to practice these skills via discussions and written assignments.

The abundant talking and thinking practice of philosophy courses transfers to other skills. *The Wall Street Journal* reports that students receiving a BA in philosophy do well on the job market, with higher earnings over time than students from many other majors. Philosophy majors also tend to earn high scores on entrance exams for graduate degrees in business, law school, and other programs. The irony here is that some parents and administrators have regarded philosophy as a worthless major that doesn't train students in a marketable skill. Quite the contrary: All the thinking and talking in the major end up minting highly marketable college graduates who have skills that can be applied to many other areas.

If you've ever seen movies such as *Legally Blonde* or *The Paper Chase*, you know that law school instructors routinely use the Socratic method, which requires students to answer questions in class. Typically, students don't know when they'll be called on, creating pressure to be up on the reading and be ready to talk on any day. Some questions require recalling information in order to answer, as when the professor asks a student to report the details of a law case in the week's assigned reading. Other questions go further, requiring reflection or thinking through hypothetical scenarios in order to reply. This pedagogical approach obviously captures all the components of asking and answering questions we discussed above.

The Discussion Project is the brainchild of Diana Hess and colleagues at the School of Education at the University of Wisconsin–Madison, my own university. The project has the tagline "Learn to Discuss and Discuss to Learn." It trains college instructors to promote better student discussions in their classes, and the project is just beginning to extend training to middle and high school teachers. The project's perspective is that meaningful class discussion requires real preparation and training, for both instructors and students, and that improved discussions are teachable skills.

The Discussion Project curriculum for teachers pushes back against the understanding-equals-learning myth to argue that discussion itself is an important learning environment. Instructors learn how to pick good discussion topics, for example by avoiding ones with a single correct answer. They also learn when to intervene versus not intervene in student discussions. Importantly, instructors also learn that students, who likely have had little experience with talking in classrooms, naturally need explicit instruction and practice to develop the skills that lead to fruitful exchanges. Lessons for the students include how to advocate for one's own position with evidence, the need to listen carefully to alternative positions, and how to cope with the awkwardness of disagreement and potential polarization in the classroom.

The research arm of the Discussion Project includes detailed assessments about the extent to which both instructors and their students gain skills as a function of the instructors' participation in the training. Instructors are strongly enthusiastic about what they've learned. Diana Hess, the founder of the project, told me that early assessments of student learning are encouraging, though the full analyses were not yet complete as of this writing.

A different approach to talking in the curriculum comes from Simon Fraser University in Vancouver, Canada. The university strongly emphasizes community engagement in its institutional mission. The Morris J. Wosk Centre for Dialogue is housed at the university, providing the space and expertise necessary to engage the community in discussions of important social issues. A natural next step for SFU was to design an undergraduate course that built on the goals and resources of the Centre for Dialogue. Mark Winston, a Simon Fraser biology professor with a keen interest in engaging with the community, developed a course called the Semester in Dialogue.

The Semester in Dialogue course embodies desirable difficulty in so many ways. It is a full-time educational commitment, and students take no other courses in that semester. Each term, the course is focused on a single complex topic such as climate change or the future of health care. Many guest experts visit the class, and ahead of each visit, students research the guest's topic and perspective, prepare questions, and make a plan for having a dialogue with the guest and with each other. Some writing assignments require students' own reflections, engaging them in valuable written self-talk of the sort we saw in chapter 4. Other assignments are aimed at engaging the community, such as writing newspaper op-eds or presentations for community groups on the semester's topic. Throughout, the students learn to listen carefully, articulate their own views, and engage in meaningful exchanges with diverse audiences inside and outside of the university. Students report that the course and the skills they gain are life-changing.

Both the Discussion Project and the Semester in Dialogue view extended time spent talking, in all its forms, as time well spent in and out of the classroom. While we don't yet have full data about the consequences of engaging in these specific programs, other research makes it reasonable to assume that students get real benefits from talk-focused activities. Like all curricular decisions, however, these projects have opportunity costs—while students spend time doing a deep dive into some topics and learn to discuss nuanced issues, they are not using that time to learn other information. Is it beneficial to spend so much time learning to discuss?

I think that there are several ways to view this question. First, we can consider what other topics students might be learning if they weren't engaging in so much discussion. For example, students in the Semester for Dialogue take only this one course in that semester. Aren't they missing out on learning from other courses? I think that this concern is playing into the understanding-equals-learning myth; those other courses they might have taken would show up on their transcript, but we don't know whether they would in fact lead to much long-term learning. Students seem to be doing quite the opposite of missing out by engaging in programs allowing them to learn much more deeply about certain topics.

Second, these programs' emphases on discussing, writing, and reflecting are building important life skills that can be brought to bear on other topics well beyond the college experience. We've seen that writing provides critical benefits for mental health as well as clarity and focus about difficult issues. These skills prepare students for graduate studies, for those who want them, and they are also highly valued by employers, as shown by the data about philosophy majors, who also actively engage in a talking, writing, and reflecting curriculum.

In other words, time spent talking while learning is not a waste of time. It's a path to remembering the material presented in class. Just as we don't want a vacation that we don't remember, we want an educational experience that creates lasting memories. Talking practice will promote that kind of learning and, in addition, leads to improved well-being and to marketable skills after schooling.

Even as I'm excited about these new programs that fully embrace talking in learning, I am hoping for more. In some active-learning programs, talking seems underutilized, because many educators may see it only as a vehicle for some active-learning exercise, not as an engine of learning itself. I hope that educators will soon embrace the learning boost that talking provides.

Talking for learning is also crucial in the next chapter, though we won't be thinking about classrooms. Instead, we'll see how talking is a route to maintaining mental sharpness in old age.

CHAPTER 6

Talking and Aging Well

"Are you listening?" Kazu continued. "When you return to the past, you must drink the entire cup before the coffee goes cold."

TOSHIKAZU KAWAGUCHI, *BEFORE THE COFFEE GETS COLD,* TRANSLATED BY GEOFFREY TROUSSELOT

IN FELIX SALTEN'S 1923 NOVEL *BAMBI* (THE INSPIRATION FOR A very different treatment in the famous Disney movie), there's a marvelous little chapter reporting a conversation between two leaves on a tree. They used to have endless leaf neighbors to talk to, but now that winter is coming, they're practically the only two left on their branch. They just can't get over how things aren't the same as in the warm summer days, and they fear that their own grip on the branch is not as strong as it once was. They try to encourage each other, but they're struggling to know what to say. After a long pause, one leaf begins to talk again, but a cold, damp wind detaches him from the branch in mid-sentence, and the conversation ends.

Besides being an engaging story about life in the forest, Salten's

Bambi is commonly taken as an allegory of the increasing persecution of Jews in Europe in the 1920s. The story of the leaves, who find life increasingly difficult and their branch-mates mysteriously disappearing, certainly contributes to that view. The leaves' predicament is also a metaphor for topics in this chapter, about both the importance of conversation in old age and the difficulties of maintaining the social bonds that enable continued talking and conversation.

We'll discuss how talking and memory change as we grow older as well as how continued talking and conversation can help keep our minds sharp as we age. It's all the talking benefits we've seen before, now applied in another stage of life. And similar to the barriers to talking in schools described in chapter 5, in this chapter we'll see that older adults' change of circumstances creates all sorts of obstacles to continued conversation and talking. We'll discuss what can be done about the decline of conversation in old age, including how we might create policies that boost talking for the benefit of older adults, their families, and society in general.

The Promise and Worry of Aging

Aging is a fact of life; you've been doing it since the moment you were born. Eventually, we get to the third act toward the end of our lives, and we'd like this act to have some verve, not drag along until the curtain comes down. Gerontologists—those who study the aging process—use the term *healthy aging* to describe this ideal outcome. Folks experiencing healthy aging have slowed down both mentally and physically to some degree, but they are still mostly independent and active with meaningful lives, and any health concerns can be managed pretty well.

Unfortunately, not everyone gets this golden ticket, and there are many deviations from the healthy-aging script. The one that older adults worry most about is losing their cognitive abilities and memory to dementia, a progressive and severe decline in memory and thinking. We're going to discuss that unfortunate outcome here, but we won't just worry about it. Instead, we'll see some good news, including how talking can help us avoid this scary ending.

Because we're tending to live longer than in the past, the number of people with some form of cognitive impairment is rapidly increasing, and those who have cognitive decline are living longer in this impaired state than in the past. Researchers estimate that about twelve million people in the United States have mild cognitive impairment, most of whom haven't received, or haven't understood, an official diagnosis. Don't be misled by the word *mild* here; this is not the normal slowing of healthy aging. Memory and thinking become noticeably more challenged, and folks become more easily confused. Living independently becomes difficult or impossible.

Mild cognitive impairment is often the on-ramp to dementia, a more severe cognitive disability caused by Alzheimer's disease or another neurodegenerative disease. Dementia patients need help with even the most basic daily activities. Eventually, as their disease progresses, they may not recognize their loved ones or their surroundings. It can be heartbreaking for them and their families.

Fortunately, it's not all bad news. Although the population boom in older adults is leading to an increase in the overall number of individuals with dementia, the chance that any particular person will be diagnosed with cognitive impairment or dementia is actually declining. Your odds are therefore improving that you'll avoid this outcome. And more good news: It turns out that at least 40 percent of dementias are avoidable, and probably others can be delayed.

Researchers have uncovered about a dozen protective factors that reduce the likelihood of getting dementia. Most of them are at least partly under our control, including components of a healthy lifestyle—good diet, exercise, avoiding smoking and alcohol. You've probably heard that these kinds of behaviors improve longevity, but you may not have realized that they also tend to improve the quality of life in old age, including avoiding or delaying dementia. And here's another thing you probably didn't know: This same research identifies another clump of behaviors that are different from the standard diet and exercise ones, but which are also associated with lower rates of dementia. These other protective factors involve talking.

Talking in Our Later Years

We'll get to those dementia-fighting talking factors soon, but our first step is understanding how talking itself changes as we age. If we're going to ask our talking to help keep us sharp and healthy, we need to understand what talking is like later in life.

Alas, even in healthy aging, we get worse at almost everything over time. Some abilities, like our quick reflexes, start to decline around age eighteen. Others start to slide in our thirties. You might think that everything is on a relentless downward trend, but that's not true. Some aspects of talking do decline, but others get better, reflecting the accumulated benefits of a lifetime of talking and conversation.

Vocabulary is a major bright spot in our talking lives, because it's quite a bit larger in old age than in our twenties. The reason is simple—our vocabulary grows via our experiences, and as we accrue experience in conversing and reading, we learn new words. This large vocabulary helps keep our language comprehension strong in the face of challenges like poorer hearing or vision.

There's just one hitch on the vocabulary front, but it's a big one. We know more words as we age, but we also have more and more trouble finding them in our long-term memory. The idea-to-word conversion process turns out to be one of the hardest parts of talking, and it gets increasingly glitchy as we get older.

The common belief is that these word-finding difficulties are a signal of old age, but in fact everyone—children, teenagers, adults of any age—gets stuck trying to find a word sometimes. Word-finding problems increase at a modest but steady state throughout our adult lives. They don't make a big jump at old age; even thirty-year-olds have trouble finding words more than people in their twenties. That's probably much earlier than you thought! The slow creep of word-finding failures is initially so subtle that early on, no one notices that they're accumulating.

What's your most embarrassing episode in which you couldn't find a word or someone's name? I'll go first. Once in a conversation about woodpeckers drilling holes in someone's house, I couldn't retrieve the word *gutter*. To keep going in what I was saying, I substituted an alternative: *water conduit*. Everyone thought this was hilarious. *Gutter* is not the most common of words, so it's not completely surprising that I couldn't retrieve it from memory, but how I managed to fail on *gutter* and yet find *conduit* is anyone's guess. Still cringey.

Younger folks who can't find a word can fend off embarrassment and suggest that they're tired, or distracted, or hungover, ha ha ha. It's a plausible excuse, because talking, including word finding, is hard work and is tripped up by fatigue or distraction. Laughing off word-finding problems becomes harder for older folks to pull off, because word-finding problems become noticeably more frequent in older adults, even with healthy aging.

There are a couple of different theories for why it grows harder to find words as we age. One is that our big vocabulary is to blame. The more words we learn over our lifetime, the more words there are to dig through to find the one we want. When we're older, we're sorting through a monstrous haystack of the fifty thousand to a hundred thousand words we know, whereas we had a much smaller pile to look through in our lower-vocabulary youth.

There is some evidence linking vocabulary size and trouble finding words. Sometimes when we learn a new word, it's a very near synonym for a word we already know. As a child, you probably knew the word *rude*, but now you also know a more grown-up word with a very similar meaning, *impolite*. There are many of these synonyms in every language in the world. In English we have *sofa/couch, pail/bucket, jacket/coat,* and thousands more. A large vocabulary is valuable in many ways, but it turns out that knowing two words for a concept can slow down our talking. For example, it takes extra effort to settle on whether to call something a *couch* or a *sofa*, compared to talking about a table, which has only one obvious word to describe it.

The extra work needed to sort through synonyms adds less than a second to the time it takes to come up with a word. This tiny delay is likely not noticeable in conversation, but the fact that it can be measured in careful studies shows that the size of our vocabulary really does affect how we retrieve words from memory to use them in talking.

We can see similar effects in bilinguals, who can talk in two languages. When you think about this feat, it's obvious that bilinguals must have truly enormous vocabularies, as they have an adult-size haystack of words for each language they know. And knowing two languages' worth of vocabulary does lead to bilinguals having slightly more difficulty finding the words they need, in both their dominant language and their less-used one, compared to people who know only one language. Again, the effect is subtle and probably not noticeable in everyday conversation, but it does show that knowing more words does affect the ease of getting them out of long-term memory and ready to be used in talking.

Another explanation for word-finding problems points to changes in executive function, the ability to focus our attention on goals and resist distractions. These abilities also start sliding in our thirties or forties. When we don't focus as effectively on our ideas as we used to, it's harder for us to get words we need out of memory. For example, adults with ADHD (attention deficit hyperactivity disorder), which impairs executive function, do have more disfluencies (*ums*, repetitions, backtracking, etc.) than those without ADHD.

The decline in executive function in old age is likely related to another change in older adults' talking: staying on topic. Older adults who score more poorly on tests of executive function tend to veer off topic in conversations more often than those who have retained more executive function skills. As we've seen throughout this book, talking affects, and is affected by, many other aspects of our brain function.

Faced with increased difficulty in finding the right words, older adults resort to some of the good-enough talking strategies that we've seen earlier. They substitute easy words for words they can't get to in the moment. For example, older adults often have trouble coming up with people's names, and sometimes they resort to saying *what's her name*, or simply *she*, even if they haven't established who *she* refers to. We can think of these word substitutions as perhaps extreme good-enough talking, to the point that they're often too confusing to the audience. These strategies may not actually be good enough anymore.

Alas, Poor Memory, I Knew It Well

When my grandmother Pearl visited me in Los Angeles many decades ago, the jacaranda trees were in full bloom. Pearl was a lifelong gardener with an artistic eye, and she was really taken with the explosion of pale purple jacaranda flowers. I drove her down street after street lined with jacarandas, and she exclaimed at their beauty and tried mightily to learn the name of this remarkable tree. She had learned hundreds of plant names over her lifetime, but now that she was in her eighties, the word *jacaranda* was just some jumble of syllables to her. Try as she might, she couldn't add just one more plant name to her memory.

Pearl's experience is exactly what older adults are worried about—the inability to learn and remember as they once could. They walk into another room and don't remember what they came in there for. They have trouble getting words they know out of their long-term memory. They can't learn people's names, new words, and other information and solidify it in their memories. In the previous chapter, we discussed the fact that learning doesn't usually happen without some effort and engagement, and now we see that the level of effort that might have been sufficient for learning when we were younger won't be adequate when we're older, even in healthy aging.

As with other aspects of cognition, memory declines slowly throughout our adult lives; we don't begin to notice the change until the memory lapses pile up. It's normal to wonder whether such lapses are the leading edge of dementia. Here's one way to think about the issue.

Increasing problems with talking and memory, along with a diagnosis of mild cognitive impairment, can indeed be signaling a path to dementia. But some amount of memory decline is present in *every* elderly person, most of whom never get dementia. The degree of memory slippage varies widely from one person to the next. Some of that variation is genetic, but other causes are environmental—effects from our own behaviors and the environment we live in. The assumption shouldn't be that there will be no decline in memory as we age. Instead, knowing that some slippage is inevitable, the goal should be to keep that slide as slow as possible.

We'll talk about two general approaches to staying sharp as we age. The first approach is treatment for those who are concerned about loss of memory or mental sharpness. This might be a change in diet, medication, or lifestyle. The goal of this kind of intervention is to keep memory and thinking stable in old age, or possibly even reverse some declines.

There are a few medications available to treat memory and cognition problems in patients who already have dementia, with new medications likely on the way. But there probably won't be medical treatments anytime soon for the gradual memory decline of typical healthy aging. Many folks have heard about ginkgo biloba or other nutritional supplements that are said to help memory, but to date there is not solid evidence that they can delay or prevent cognitive impairments.

The second approach is prevention, begun prior to any noticeable declines in cognition. A huge idea in the field of gerontology is that some people seem to have a large *cognitive reserve* that helps them weather the effects of aging and disease, while others do not have this cushion. Those with cognitive reserve seem to maintain their mental capacities longer over time. And if they do eventually succumb to a neurological disease such as Alzheimer's, they decline more slowly, compared to others with less cognitive reserve. Folks with and without a large cognitive reserve likely differ in their genetic makeup, but the big news is about how our behavior and experiences can build up a cognitive reserve well before we get to the challenges of old age.

One way to build up cognitive reserve is for older adults to exercise their brains more. A talking-related exercise that is sometimes linked to improvements in healthy aging is being bilingual. Bilinguals who use different languages in different situations have, in addition to all the normal difficulty of talking, the added burden of selecting one language and inhibiting the other. The idea is that this language-switching activity provides an executive function brain workout, a desirable difficulty that leads to better executive function abilities in old age. Does it work? Well, I wish I knew. There have been hundreds of studies addressing this question, but no clear answer. Some studies show an advantage in bilinguals' executive function, some show a disadvantage, and many show no difference compared to monolinguals. My feeling about the conflicting results is that talking is a substantial desirable difficulty and executive function exercise on its own, and the bilinguals' extra burden of language switching doesn't necessarily add much additional benefit.

For bilinguals or monolinguals, other opportunities for brain exercise and building cognitive reserve might come from playing cognitively challenging mental games. There are now all sorts of games to play on computers and tablets and smartphones, often called brain-training games. Most of these programs have a cluster of different games, each only a minute or two long. In each little game, players see or hear some information and then must do something with it—recall it later, judge whether it fits a pattern, select some images that follow a rule, and so on. Most of the games are timed, and both speed and accuracy count in the player's score. The games resemble some diagnostic tests you'd find in a neurologist's office, but with colorful displays and encouraging feedback to keep people playing.

Throughout this book, I've been talking about the value of experience and practice. These memory-training games do give players practice recalling information, keeping focused, switching from one task to another. You might therefore think that I'd be excited about these brain-training games, that I'd recommend them. But no, I am not a fan.

To understand why I and many other researchers are not sold on these brain-training games, we need to understand what kind of practice does and doesn't improve memory, executive function, and everyday thinking. Imagine for a minute that we gave a group of elderly adults some squeezy balls for hand exercises. We tell them to do the exercises daily. After a month, we come back to see how they're doing. We might expect that folks' grip strength would improve from exercising with the balls. Maybe some of them would report practical benefits, like an improved ability to twist off caps and jar tops.

Would their memory improve? No, you say, that's silly. There's no obvious link between ball squeezing and memory. Quite true. Now imagine that we repeated the study with a new group of older folks but now with more-colorful balls that we call memory balls, with a picture of a brain printed on the side. What happens after a month of hand exercises with the memory balls? Well, maybe some people would experience a placebo effect—they might think their memory is better because they did their memory ball exercises. But in reality, the same thing happens as in the first study—exercising with a squeezy balls, no matter what you call them, does not affect anyone's memory.

I'm telling you this story because I think that many brain-training games are not much different from the memory balls—they have the word *brain*, or *memory*, or *cognition* in their names or advertising pitches, but that does not mean that playing the games will improve anyone's memory or cognition. Many investigations of these games have found their effects to be very limited—basically, the players get better at playing the games, but that success does not transfer to any meaningful improvement in everyday life. Definitely, you should play these games if you enjoy them, but don't expect a boost in everyday memory or thinking.

Why don't the games translate to real life? Our story begins with the fact that memory researchers like to distinguish among different kinds of memory—long-term memory, short-term (or working) memory, memory for words, memory for episodes in our lives, memory for how to do stuff like ride a bike, and so on. This division into different memory types is useful for researchers because it gives them specialized vocabulary to discuss research and theories of how the subcomponents fit together. A potential downside of all this partitioning, however, is that naming all these separate pieces can increase researchers' tendency to view each subcomponent as an independent entity and separate from other aspects of memory and cognition.

That's a problem, because even though these memory subcomponents have their own names in research discussions, they are well-connected in real life. Those connections affect how memory works in old age. Remember my grandmother Pearl and her struggle to learn the word *jacaranda*? Her repeated memory failure, and later one small success that I haven't yet told you about, provide a potent example of these connections in older adults.

Pearl's problem was that the word *jacaranda* wasn't anchored to anything else she knew. If jacarandas were instead called purple-blossom trees, she would have remembered the name, because the words *purple*, *blossom*, and *tree* were familiar to her, and together they had a meaningful connection to what she was seeing. Without a familiar concept or word to cling to, she was stuck, unable to learn. But then she had a breakthrough. Once when repeating the word *jacaranda* after me, she said, "Oh, it sounds like *jacket veranda*!" Suddenly, problem solved, and from that moment on, she talked about the jacket veranda trees. She knew it wasn't quite right, but it was absolutely good-enough talking. I knew what she meant, and she was pleased to have outsmarted that unlearnable *jacaranda* word.

There are several important points here. The first is, whoa, *veranda*?? Score one for older adults' massive vocabularies and how they can put all those words to good use. The second point is the connectedness of memory. Pearl couldn't learn a new unfamiliar mashup of syllables, but she could make progress when she drew on words she already knew. Older adults' vocabularies and other long-term memories, grown over their long lifetimes, are essential support for their weakening short-term memory and learning.

The brain-training programs ignore the importance of old knowledge for new memories. Indeed, most of these programs deliberately avoid any connection to what people already know. The games typically have only disjointed new information, such as a handful of random numbers, nonsense squiggles or dot patterns, or meaningless combinations of syllables. The games deliberately downplay connection to real life and knowledge and as a result, they don't bring real-life benefits.

From all this we can see that for older adults to keep their memory, executive function, and attention sharp, they need practice that connects new information with what they already know. What kind of activity connects old knowledge to new and also is a desirable difficulty in exercising the brain? Yeah, you know. Talking.

Talking as Brain Training

The idea that talking is an important brain-training exercise fits very well with the dementia researchers' recommendations I mentioned at the start of this chapter, the ones that are linked to staving off cognitive decline and dementia. Beyond a healthy diet, exercise, and other protections from dementia mentioned previously, these researchers found that several talking-related factors are associated with lower levels of dementia. One is having an active social network, which has social and emotional benefits and also promotes talking. Another is addressing hearing loss with up-to-date hearing aid settings, which obviously helps older adults maintain these social interaction and talking benefits. Others are avoiding brain injury via falls or strokes, and continuing group physical activity. All of these have their own health benefits but also promote folks getting together, socializing, and having conversations.

Talking creates an important practice that strengthens memories. Every time we recall events from our past in order to talk about them, we strengthen both the memory of the event and the words we're using to talk about it. Talking is brain training that naturally is integrated with what we already know and what we already do in the real world.

Talking also aids attention and executive function. We previously saw that both little kids and younger adults benefit from self-talk to get themselves in the mindset to concentrate. There has been relatively little extension of this work to elderly adults, but the available evidence is encouraging. Self-talk does appear to help older adults compensate for declining abilities to focus. Though more research is needed, it's possible that getting in the habit of naming the thing you want from the next room could help you remember what you had gone in that room for when you get there.

Beyond talking to ourselves, we've seen that conversation with others has special desirable difficulties that should be extremely beneficial for maintaining memory and cognition. Conversation has timing demands that keep all the conversation partners on their toes, focusing their attention. The range of topics keeps each person reacting to new information, relating it to what they already know. And each person in the conversation must manage several jobs at once—all the subcomponents of talking, but also comprehending, planning what to say next, predicting when the talker will finish, and so on. Conversation partners switch back and forth between the talker and comprehender roles, naturally leading to practicing task switching, a component of executive function.

All of this prior research on the desirable difficulty of talking forms the basic rationale for why conversation and other forms of talking ought to protect our cognition in old age. But now we need to get to the hard data, specifically for elderly adults. The dementia researchers' recommendations for talking-related activities stem from their search for associations between behaviors across thousands of research participants answering survey questions about their lives, such as the size of their social network. This correlational research, such as studying the relationship between dementia and diet, social activities, and so on, is carefully done and the recommendations are solid. But there's still room for studies about whether talking behaviors have a direct role in preventing dementia.

To fully see effects of talking on elderly adults' cognition, I'd like to see studies that conduct a talking intervention. In this kind of research, elderly adults are randomly assigned either to a group that has some intervention that adds extra conversations or other forms of talking to their days, or to a group with no changes to their activities. After some period of time has elapsed, the researchers check back with the two groups to see whether they differ in their memory, cognition, health, and so on. This kind of study would help us understand whether real-world policies of adding more talking to elderly adults' lives could promote healthy aging and protect them from dementia.

There are vanishingly few talking intervention studies like this, in contrast to endless studies of the effects of brain-training games. You, who are reading a book about the hidden benefits of talking, may find this situation surprising, but remember that many cognition researchers have tended to consider memory and executive function as largely divorced from talking. As a result, these researchers probably haven't thought to investigate the effects of talking on memory and executive function in elderly adults.

After some digging, I did find some intervention studies in elderly adults comparing an increased-conversation group to a control group with no changes in their activities. In one study, college student volunteers visited nursing home residents twice a week for about two months. The volunteers conversed with residents for about an hour on each visit, and in one group of participants, they also played card or board games with them. At the end of the two months, the researchers found that the residents who had these extra conversation visits had improved memory, quality of life, and engagement in the daily activities in the home. A control group of residents in the same homes, who had no conversation visits, showed no improvements.

A much more ambitious study targeted a large group of elderly adults living alone who had diagnoses of depression and mild cognitive impairment. The researchers tried a huge talking intervention: Five days a week for three solid years, the elderly research participants in the conversation group had a member of the research team stop by for an hour of conversation. The emphasis in these visits was on the older adults' own talking—the researcher asked questions and encouraged the elderly folks to reminisce about their lives and other topics. Weather and mobility permitting, the visitor and the older adult took a walk while they talked. At the end of the three years, the elderly folks in the control group, who had no conversation visits, had deteriorated in many ways. Memory and cognition had worsened, depression had increased, and quality of life had sunk. These results are not surprising; a diagnosis of mild cognitive impairment is usually followed by more deterioration in the next several years. The folks in the conversation group looked very different. They had retained their cognitive and memory skills at a steady level over the three years, their depression had eased, and their quality of life had improved. Since all the participants began with a diagnosis of mild cognitive impairment, it's quite remarkable that the ones in the conversation group held on to their cognitive abilities so well over this time.

Both of these studies show that engaging older adults in more conversation has real benefits for their memory, cognition, and quality of life. We can't know exactly how much of the benefits had come from the social aspects of the visits—a friendly face comes by to see an elderly adult—and how much was from the cognitive benefits of talking. But as I've shown, talking interacts with many other skills we have. This proof of the benefit of talking interventions is really important news with practical applications, given the many burdens of dementia for the patients, their families, and society at large.

Though these talking interventions do slow cognitive decline, the hitch for wider adoption of programs like these is the effort needed to deliver them. Increasing conversation via individual visits to each elderly person is both valuable and enormously labor-intensive. It couldn't be scaled up to cover every elderly adult in need. To have a wider reach, we need to think about how else to promote the benefits of talking in old age.

One option is to look to technology to help bring conversation to isolated elderly adults. Some studies have used software that prompts dementia patients to remember past events and talk about them with a caregiver. Another project connects elderly adults with a conversation partner remotely, via video calls. Many of these projects are in their early stages, and results aren't necessarily complete at this point. They also continue to be labor-intensive, requiring human conversation partners in real time, even if they're connected remotely.

We might instead encourage elderly adults to have conversations with conversational agents—artificial intelligence systems that simulate human conversation partners. You're likely already familiar with simple conversational agents like Alexa or Siri, who can respond to requests, and you may have heard about or conversed with ChatGPT or other artificial intelligence systems. For example, an elderly person might speak with an image on a screen of a human-looking agent who replies with fairly natural vocabulary, intonation, facial expressions, and gestures. Conversational agents can be specifically designed for elderly adults, tuned to topics and conversational styles that older folks might like. They might be good at getting older adults to recall the past and talk more, with benefits for their memories. These agents' capabilities are rapidly expanding, and researchers are investigating whether elderly adults would find them interesting and would engage in conversation with them. The tentative answer is yes, though more work needs to be done.

An option that's already commercially available is animatronic pets for elderly adults. The "pets" don't talk themselves, but they can encourage talking and social interaction. When my mother had a diagnosis of mild cognitive impairment and had very little mobility,

I gave her a battery-powered cat that meowed, purred, and washed its face with a paw. Oh, she loved that cat. She knew it wasn't real, but she wrapped it in a blanket, talked and sang to it, and told me what a clever toy it was. "It's like I'm having a second childhood!" she said, a comment that was so perceptive, it brought tears to my eyes.

Even though the cat didn't talk back, it continued to be a focus of talking for my mother and a conversation starter to encourage talking with other people. Of course, others' reactions could be different, and I don't know of any formal studies of the effects of toys like this.

Although many of these forms of technology are not fully available now, they will likely be essential components of elder care and social enrichment in the near future, given the increasing numbers of dementia patients as populations age. To date, the implementations haven't all been successful. Japan has invested heavily in cheerful talking robots in nursing homes, but with mixed results. Instead of enriching residents' lives, the introduction of robots sometimes reduces the amount of conversation that residents have, because staff now needs to deal with the glitchy robots, with less time to talk to residents. Other robot implementations have met with better success.

Technology will probably continue to be helpful, but we should also seek low-tech solutions. Why don't we encourage elderly adults to talk more to each other? You may be rolling your eyes at how utterly obvious this idea seems, but practically speaking, it may be quite difficult to implement. There are all sorts of possible routes to more talking, but also all sorts of real-world barriers to making these plans effective for elderly adults.

To get a sense of both the options and the complications, let's imagine Roy, seventy-eight years old, living with his wife, Bernice, seventy-six, in the same house they've lived in for decades. Both have been retired from their jobs for about ten years.

When Roy and Bernice retired, their social circles and talking opportunities instantly shrank. That's typical, because practically every retiree leaves behind colleagues and conversation topics associated with work. Retirees may be delighted to let go of irritants in their work lives, like inflexible rules, tedious meetings, or an idiot boss. It doesn't necessarily occur to them that they're also leaving behind extensive social interaction that's important for mental sharpness. Remember the studies about how our vocabulary increases with age? Well, the fine print is that often vocabulary increases until about age sixty-five, when it levels off or declines. I don't know of any research to back up this hunch, but I have to wonder whether this leveling off at sixty-five is at least in part from the reduced interaction and stimulation that comes with retirement around that time.

These days, most of Bernice's and Roy's conversations are with each other, but they do have some additional social networks. They talk on the phone with their kids and grandkids about once a week and visit them a few times a year. Roy has casual chats with other dog owners when he's out walking Betsy, and he gets together with some old friends for coffee most Tuesday mornings. Bernice does a weekly lunch with some girlfriends. Sometimes she and Roy see another couple for dinner or go to a show.

Roy and Bernice are not isolated or lonely, but there's still a lot of room for more talking and socializing in their lives. Moreover, as they age, they're each encountering continued narrowing of their social networks. Friends move away, travel, become less mobile, all resulting in fewer people to talk to. And Roy is now illustrating one of the important talk-related factors that those dementia researchers identified for retaining mental sharpness: the importance of good hearing aids, updated regularly. Roy hasn't realized it yet, but he needs to have his hearing aids adjusted. He isn't following conversations as closely as he used to, and so he's engaging and talking less when he does get together with friends.

Sometimes arthritis pains keep Bernice from going out, so she misses lunches or other excursions. She and Roy are beginning to think about moving to someplace with no stairs and less upkeep. They've thought about an apartment nearby, or a senior independent living situation, or moving away to be closer to the kids and grandkids. The many choices and their costs and benefits feel overwhelming, and they haven't made any decisions.

Bernice and Roy could use some help thinking through these complex future paths, because they, like most older adults, have more difficulty deciding among options than they used to. Elderly adults like Bernice and Roy tend to stick with their current situation when faced with many other options, even when a change is likely beneficial, such as letting go of a high-maintenance two-story house. And Roy and Bernice could benefit from more talking, and talking with a greater number of conversation partners, to compensate for the narrowing options in their social circle.

How do we make that happen? Some of these suggestions involve getting out and trying new social activities, while others are more solitary pursuits that can be done at home. These home-based ones don't provide the stimulation of conversation, but they likely still have some talking-related benefits.

- Game get-togethers. Bernice and Roy could play games with their friends. The local senior center probably has scheduled communal games like bridge, dominoes, mah-jongg, Scrabble. Card and board games of all sorts promote talking, decision-making, cognition, and social bonding.

- Senior centers often host other stimulating activities, including talks, lunches, movies, musical performances, exercise groups, and classes on art, woodcarving, and so on. I've given talks about talking and aging at several senior centers, and even the smallest ones in tiny villages are crammed with weekly activities that bring people into the center and get them interacting.

- Reminiscence writing programs have older adults write short reflections about important times in their lives. As we've seen before, writing and recalling the past can lead to more insight and well-being. Writing down reminiscences also provides wonderful opportunities for older adults to talk, recall information, and strengthen their memories. The written stories are often very interesting for other family members and can be a great conversation starter.

- Continuing education classes at a local community college or university are excellent for older adults. Learning something new is extremely stimulating, and talking about this new topic broadens conversations for both the learner and their conversation partners.

- Book groups, which involve reading something new and talking about it, are good for the same reasons.

- Volunteering can be inherently meaningful and creates a structure in which to meet new people with shared goals, as well as new information and conversation topics.
- Crossword puzzles put older adults' large vocabularies to use and provide excellent memory and talking-like stimulation, with proven benefits for talking and executive function skills.
- Other word games haven't been studied extensively and don't necessarily have proven benefits, but it is plausible that at least some of them could be beneficial. Word and language games, such as Scrabble and Boggle, are available in computer, phone, and tabletop versions.
- Twenty questions and similar games encourage the player to combine cues to guess words, which is similar to retrieving words from memory for talking.

Which of these do you think Bernice and Roy might do? Maybe if their kids set up some new games on their phones they'd play them, but I'm guessing that suggestions for new clubs or social interactions would fall flat. Bernice and Roy seem pretty set in their lanes. Like many other elderly adults, they're more likely to continue with the status quo than to try something new.

And remember, talking is hard work, and the prospect of meeting a bunch of new people and talking to them may be a deterrent to trying new activities. Especially if it's getting harder for Roy to hear or if Bernice is worried that there won't be a nearby parking spot at the senior center when her arthritis is acting up. Older adults tend to prioritize existing social relationships over making new friends, compared to the tendencies of younger adults. This reluctance to make new friends probably stems from many forces, but I can't help thinking that one of them is aversion to the hard work and potential awkwardness of making conversation with new people, compared to the ease of familiar routines.

You and I might believe that Thursday board games at the senior center would be fun and an excellent desirable difficulty for Roy and Bernice's social interaction and talking. But they quite naturally may have different ideas. They may decide to just skip the whole thing and sit at home, not talking, watching TV.

I'm not saying that all elderly adults are refuseniks like Roy and Bernice. I've known several retirees, mostly ones who moved to a new city to be near children and grandchildren, who threw themselves into their new environment and found classes, volunteer work, and new social circles that enriched their lives. Nonetheless, it's important to realize that the dominant reaction is likely to be resistance when older adults' doctors, therapists, or children push them to disrupt the status quo and increase their social interactions.

Here's where the idea of cognitive reserve comes in. Building up our talking, social interaction, and learning earlier in life strengthens our memory and cognition. It builds up a protective stockpile of skills that can help us later as we naturally slow down in mind and body, and when our social circles narrow. Researchers typically think of cognitive reserve as a lifelong accumulation, gleaned from education, having a stimulating job or hobbies, and so on. A lifetime of stimulating activities is indeed valuable, but I don't believe we should limit our thinking to only a whole-life timescale. After all, we've seen the value of increasing conversation in interventions that were only a few months or years in duration.

With that point in mind, here are some suggestions for building cognitive reserve and combating the inevitable narrowing of social opportunities in old age. We need a concerted societal effort to increase social networks and talking at a point when adults are particularly receptive to changing their habits. Many of these periods will be around other big life changes—a new job, moving to a new town, children moving out of the house, and the really big one: their own or a partner's retirement. Starting to increase cognitive reserve at the time of retirement is not a substitute for a lifetime of stimulating activities, nor is it a substitute for later interventions. However, folks who are contemplating retirement may be particularly open to new activities, including ones that boost social interaction, talking, and a cognitive reserve. To paraphrase the quote at the beginning of this chapter, let's reach these folks before the coffee goes cold.

There are already plenty of news articles and websites and listicles with headlines like "85 Things to Try in Retirement." These are great resources, but I'd like to see us find ways to go beyond information sharing. I'm not at all an expert on how to find new retirees or create behavior change in this group, but I can think of more knowledgeable folks who could contribute. For example, public health researchers routinely study how to persuade people to develop healthy behaviors, such as quitting smoking. These researchers could similarly investigate how to promote the social interaction that is known to be protective against dementia. They could investigate when to target people, how to find them, what kinds of messages are most persuasive, how to follow up, and so on. Given the economic and social costs of dementia care, success here could have important consequences for public health and limiting health care costs.

Health insurance companies also have a vested interest in keeping their enrollees healthy, including maintaining their cognitive health. The occasion of their customers' retirement or their adult children leaving their health insurance could be excellent times for health plans to offer coupons, appointments with consultants, free wellness classes, or other inducements to try new activities. Similarly, pediatricians recommend reading to children and all sorts of activities to new parents in a coordinated push with the American Academy of Pediatrics; why can't we develop a similar infrastructure for gerontologists and other doctors to provide more specific recommendations and support promoting talking and social interaction for retirees? I'm sure that individual doctors make good recommendations to their elderly patients, but I'm equally sure that there's room for more here.

Many people in their sixties might think of themselves as too young for the senior center in their town, but we know that at least some of them eventually decide to check it out. Gerontologists do study the factors affecting participation in senior centers, but there is likely more room for research that specifically addresses promoting earlier engagement with the center, so that it becomes an enriching force earlier in old age. There's probably also more room for getting this research into the hands of center directors and activities directors to enable them to find new ways of outreach in their own communities. And my nonscientific survey of some continuing education programs for older adults suggests that at least some of these groups have had more success developing fantastic programs than getting the word out about them. Perhaps there's room for some continuing education programs and senior centers to band together for occasional special activities fairs where they promote their various programs and have sign-up sheets. If only a small percentage of elderly adults boosted their social interactions and talking, that would still be a win.

Some of these suggestions are likely dumb, or at least not feasible in all contexts. But the idea of concerted efforts toward increasing stimulating activities in recent retirees, before they get too set in their routines, is not at all dumb. If we try to promote this social interaction, we'll be using what we know about the benefits of talking and applying it to people who need it. Given the population surge of elderly adults and the enormous costs of cognitive decline for individuals and society, we can't afford to act like old folks and stick to the same paths we've gotten used to. Policymakers, physicians, social service providers, and family members need to help older adults increase their talking, interaction, and other behaviors that are known to protect their mental acuity, increase quality of life, and forestall some rising health care costs.

THE THREE CHAPTERS IN PART 2 HAVE FOCUSED ON HOW TALKing can tune the talker's brain and how we can think about social policies that can enable more people to benefit from talking. In part 3, I broaden the focus from individual brains to our environment. We'll see that because of the way that talking works, it has all sorts of effects on the world around us. Most of these consequences of talking are just as hidden as everything we've learned about the brain-tuning effects of talking in part 2. We start in chapter 7 by looking at how the nature of talking and talk planning shapes our own language and all the languages of the world.

“

PART THREE

Talking Out in the World

”

CHAPTER 7

Talking Changeth Language

Everybody say when you hear the call
You've got to get it underway
Word up!

CAMEO, "WORD UP!"

I'M WRITING THIS BOOK IN ENGLISH, BUT IT'S NOT THE ENGLISH of Shakespeare's time, four hundred years ago. It's not even the same English that my grandparents spoke, with their talk of dungarees and davenports and other expressions that are no longer on fleek. Even *on fleek* may not be so on fleek anymore and on the way out.

All languages are constantly on the move. It's not just that some words go out of style and other ones come in; pronunciations and grammar shift around too. Researchers who study why languages are the way they are and why they change over time often describe them in Darwinian terms: Languages evolve and adapt, improving their fitness over years, decades, and centuries. Certain features of languages—their word orders, their prefixes and suffixes, their pronunciations—turn out to be useful, and these features may become more common in that language and may spread to other languages over time. These useful language features drive out less fit alternatives. It's a form of natural selection, not for features in species like ear shape or the color of fur, but for language features like speech sounds or word orders.

This survival-of-the-fittest perspective for languages naturally leads us to interesting questions: What would it mean for a language to be fit? And who is benefiting from the increased fitness of a language? Languages are so complicated that fitness will likely include beneficial features for language learning, for comprehending, and for talking. Strangely, these three kinds of fitness haven't received equal research. When linguists think about language fitness, they most often think about the needs of young language learners, suggesting that more fit languages would be easier for children to learn. There's also some discussion of the needs of comprehenders—that a more fit language would be easier to comprehend than a less fit one. But there's comparatively little discussion of how languages might be fit because they suit the needs of talkers. I think that's a mistake. Since talking is more difficult than comprehending, making talking easier should be a powerful force in making languages more fit.

To see some ways that languages adapt to the needs of talkers, we need to not just think about our own language but also look at language patterns that we see across the globe. There are approximately seven thousand spoken languages and three hundred sign languages in the world. This chapter is about how talking shapes these languages, both in the moment and over time. We'll see some patterns that seem to have won the survival-of-the-fittest contests, and some language patterns that may be dying out. We'll consider whether the popular patterns in the world's languages are ones that benefit talking more than the rare patterns that seem to be on the decline.

How Climate Shapes Language

One pattern that is consistent across all spoken languages is that they all have both vowels and consonants in their repertoire of speech sounds. Researchers trace the universality of vowels and consonants to how our jaws work. The basic thing that jaws can do is open and close. We do the jaw open-close routine over and over when we're chewing food, and we carry this alternation over to speaking. When jaws are open during speaking, vowel sounds come out, and when they're almost closed, we get consonants. The fact that we roughly alternate vowels and consonants in our words appears to come from our alternating jaw open and closed positions.

Things get more interesting when we realize that there's a major divide in the world's languages in how much they populate their words with vowels versus consonants. Hawaiʻian, the language spoken by the indigenous people of Hawaiʻi, is all about the vowels. It has only eight consonants, and it has ways of lengthening and combining its set of five vowels to yield twenty-five different vowel sounds. You can hear Hawaiʻian's vowel-loving approach in the words that name Hawaii's islands: *Oʻahu*, *Kauaʻi*, *Maui*, *Hawaiʻi*, and so on. At the other end of the spectrum, German is a proud member of Team Consonant. You can hear how they pack in the consonants in words like *Entschuldigung*, which means "sorry," and *Angstschweiß*, meaning "cold sweat." There are thousands of other vowel-loving languages like Hawaiʻian in the world, mostly with speakers living in tropical locations. And there are thousands of other consonant-loving languages like German, with speakers in nontropical areas. Why?

The answer to this question was a mystery until quite recently, when language researchers thought about the effects of dry versus humid air on talking. As singers surely know, it's more difficult to sing when the air is dry. That's because it's harder to make our vocal cords vibrate, and we need that vibration for singing. We also need that vibration for producing vowels. This fact makes vowel-heavy languages well suited for humid climates like Hawai'i's. But in dry climates like deserts and cold, dry areas in parts of Germany, talkers can reduce talking effort by using their vocal cords less, and as a result, producing fewer vowels and more consonants. Dry air influences talking difficulty, and each language's vowel and consonant usage tends to be well suited to its climate. Although there are not historical records going back centuries for many of these languages, researchers have hypothesized that a language's consonant/vowel mix changes gradually over time, making each language a better fit for the climate in which it's spoken. That's our first example of how talking difficulty affects languages.

In addition to the dryness of the climate, temperature also appears to affect the patterns of vowels and consonants. Vowel sounds travel well through warm air and can pass through nearby foliage, making a vowel-heavy language useful for conversation in a jungle. In colder climates with a lot of open air, consonants, and consonant groups like *st* and *pr*, are more audible than in warm jungles, making consonant-heavy languages beneficial nearer to the poles than in tropical climates. These language patterns stem from the physics of sound transmission through the air. We can see similar patterns in birdsong, with more vowel-like bird calls in warmer climates and complex warbles in cooler ones. The endnotes have some links to examples of birdsong in different climates, including the European starling from northern Europe (also an invasive species in the United States), whose song is a complex set of twitters and consonant-like clicks, and the white bellbird of the Amazon rainforest, which opens its mouth wide to belt out a single deafening note.

Some researchers think of these effects of environment on speech or birdsong as being tuned to the needs of whoever's listening. For example, birds in urban areas near automobile or construction noise sometimes change the pitch of their song, which may make their song more perceptible in the face of urban noise. On that view, languages and birdsongs end up with different sound qualities so that the listeners can hear the speech or song better, or maybe so that the producer of the talk or birdsong can hear themselves better. Definitely there's a give-and-take here between whoever is producing the sound and who's perceiving it, and we'd expect that many factors contribute to fitness. However, the fact that climate affects the effort of producing certain sounds shows that talking difficulty itself really does contribute to the fitness of a language for its environment.

Word Finding Changes Languages

The examples we just saw were about how the physics of speaking—the way the jaw moves, the way dry air restricts vocal cord vibrations, the way sound travels—affects the inventory of sounds in a language. The next examples move away from the physical act of articulation and turn to the effects of how we plan our talk. In chapter 4, we saw that the first step of talk planning, concentrating on our ideas to get words out of our long-term memory, has all sorts of consequences for our focus, executive function, and more. Now we're going to see how these very same word-finding processes change our language.

You know the expression "He chose his words carefully"? Why do we have this saying? If everyone always chose their words carefully, no one would ever need to point out situations of careful word choice. We have this expression to describe an exception—something we can do if the situation demands unusual thought or tact. We saw in chapter 3 that instead of carefully choosing the absolute best word, most of the time we maximize talking efficiency by quickly and unconsciously picking easily accessible words that are just good enough for what we want to convey.

Our good-enough word grabbing not only affects our individual word choices; it also changes our language. Before I explain how, I'd like you to do a little exercise. Look around you and think of the names for five to ten things you see. When you're done, let's examine your list. If you're inside, you probably named small objects like *phone*, *glass*, *book*, or *pen*, maybe some furniture like *chair* or *table,* and maybe some features of the general area, like *wall* or *floor.* If you're outside, maybe you said *grass* or *cloud*.

These words, and probably almost all the other ones you listed, have something in common. They are ambiguous, meaning that they have more than one meaning. The word *pen*, for example, refers to both a writing implement and an animal enclosure. And *table* refers to a surface you put your book on, but it also can refer to a chart of numbers or data, and it can be a verb in expressions like *The chair tabled the motion*. (*Chair* and *motion* are ambiguous too!)

Before I started studying talking, I studied how people interpret ambiguity in language. As our little exercise just showed, ambiguity is everywhere. Most words have several meanings, and the more common a word is, the more meanings it's likely to have. And guess what? When we're the audience for someone else talking, all this ambiguity can make us work (mostly unconsciously) to figure out the talker's intended meaning.

Many of the research articles about comprehending ambiguity, including ones I wrote myself, start out in basically the same way: How can people possibly understand language so well, given how much of language is ambiguous? Eventually we researchers homed in on what was happening: People figure out a talker's intended meanings for ambiguous words by making rapid unconscious guesses about what the talker likely meant. These guesses take into account which word meanings are overall more common in our experience and which ones are likely most relevant in the current situation. Sensitive measures of this unconscious process of settling on the most likely interpretation of an ambiguity show that it does take more time than if there were no ambiguities, but in general our guesses are so rapid that we don't typically notice the work we've done. This bit of extra effort is tolerable because overall, the audience has an easier job than the talker and has time to do this extra work.

About the time that this story of how we handle ambiguity became clearer, I shifted my attention to studying talking. As I learned more about how talking works, I realized that there's a very different, very interesting question about ambiguity that I hadn't really appreciated when I was focusing on just the understanding side of things: If ambiguous language is harder to comprehend than unambiguous language, why is there so much of it?

The answer is that ambiguity exists because it benefits talkers, and our good-enough talking actively contributes to the rise of ambiguity in our language. Here's an example. Imagine seeing a friend pouring orange juice who's accidentally missed the glass with some of it. Your friend doesn't notice, and you say, "Hey, the juice is running off the counter!" Why did you say *running* rather than *dripping* or *spilling*? Probably *running* won out because it is a much more common word than *dripping* or *spilling* and therefore it's likely to come from memory earlier than those others, and therefore get picked in the plan for what to say.

But *running* is also an incredibly ambiguous word. Runners run, but so do clocks, cars, noses, trains, stockings, rivers, politicians, and more. The website Dictionary.com lists ninety-seven different meanings of the various forms of the verb *run*, while the verbs *drip* and *spill* have only a couple of meanings each. As we're picking words for our talking, we're usually oblivious to the possibility of other meanings, again because of the way talking works. When your brain chose *running* as the word to describe what the juice was doing, it was focused only on whether it was a good-enough match for that situation. The fact that *running* has several dozen other uses was not information that was available when the brain was focused on the current idea. So even if we wanted to avoid ambiguity, the brain's word-picking procedure for talking makes that nearly impossible to do.

It's not an accident that common words like *running* tend to be more ambiguous than rarer words like *dripping*. By continually choosing good-enough common words that come rapidly from memory, talkers extend the usage of these common words to more and more situations, making them more and more ambiguous. In this way, talking very naturally changes our language to create more ambiguity. Not just in English, but in any language where talkers are talking, picking their good-enough words.

Planning Word Order Shapes Languages

As soon as talkers make word choices during talk planning, they have to get going on assembling the words in an order. This stage of planning also has enormous effects on how the languages of the world tend to work.

I remember being stunned to learn as a kid that speakers of Japanese and German typically put verbs at the end of the sentence. My egocentric kid brain, steeped in the patterns of English, just could not fathom how anyone could possibly speak that way. These verb-at-the-end languages get the last laugh though; it turns out that there are more languages in the world like Japanese and German, with verbs at the ends of sentences, than with verbs in the middle, as in English.

This little example offers a reality check on our intuitions—just because we think some unfamiliar language pattern seems difficult doesn't mean it actually is a burden for people who grew up with it. To look beyond our intuitions to understand which language patterns might be more fit than others, we'll look at what patterns are frequent or rare across the world's languages. In the Darwinian survival of the fittest, a language with an extremely rare pattern may be less fit than languages with patterns that appear over and over again across the globe.

Most languages in the world have a favorite order for arranging their words. We can see the favorites in very simple sentences that describe one action, one actor (someone doing the action), and something being acted on. For English, the sentence *Chris drank water* shows our basic order for these three components: first comes the actor, *Chris*, then the verb *drank* conveys the action, and last is the thing acted on, *water*. Japanese, German, and many other languages like the verb at the end: *Chris water drank*.

With three words to arrange for action, actor, and thing acted on, there are six different word-order possibilities. If every ordering option was equally fit, we'd expect to find that each of the six orders is roughly equally common, with each found in about one-sixth of the world's languages.

This doesn't happen. Some of the six possibilities appear in thousands of languages, which together are spoken by billions of people. Others are vanishingly rare, occurring in only a few languages, often ones with only a few thousand people using them. Why?

To help us dig into this puzzle, I've shown the six word-order options grouped by how common they are across the world's spoken and signed languages. The world's three most common options are on the left. They're the basic orders in more than 95 percent of the world's languages, including the Japanese/German option at the top left, the English option in the middle, and a third one at the bottom of the left column, which puts the verb first. The common property of these three word orders on the left is that they place the actor before the thing acted on; in our example sentence, that's putting *Chris* somewhere before *water*. The word orders in the right column put the acted-on element, *water*, before the actor, *Chris*. These languages are extremely rare, found in fewer than 5 percent of the languages.

VERY COMMON: ***Chris* before *water*** **(95% of world's languages)**	**VERY RARE:** ***Water* before *Chris*** **(5% of world's languages)**
Chris water drank	Drank water Chris
Chris drank water	Water drank Chris
Drank Chris water	Water Chris drank

Such strong tendencies in word-order patterns across both spoken and signed languages suggest that there must be some powerful forces at work here. This one isn't about humidity or temperature. Instead, there's some force in our brains pushing us to mention actors before things that are acted on.

What's the big deal, you might be saying, it's just more natural to put the actor before what's getting acted on. It certainly seems more natural to me, but we really don't want to rely on our intuition here. We don't know if the actor-first options seem good to us because they really are somehow more natural for human brains, or whether they seem better just because they're more familiar to us. Also, even if those actor-first options really are somehow more natural for our brains, we'd still like to know why.

Here are some theories that have been suggested for why languages tend to put actors before what's acted on. One idea is that all human languages are descended from a single original language that used the same *Chris drank water* order that we use in English, and they branched out from there. In this hypothesis, the three options in the left column are popular because the world's first language started out with one of these options and all the languages that descended from this one didn't drift too far from the original basic pattern.

I don't find this explanation very satisfying. We don't know if all spoken languages are descended from a single parent language, but even if there was a single original language, this still doesn't explain why an original *Chris drank water* word order routinely morphed into the other word orders in the left column but almost never drifted over into the right-hand column. Another strike against this explanation of drift from a mother tongue comes from sign languages. Most sign languages arose in isolated communities and are not related to each other, yet they also have an extremely strong bias toward putting actors before acted-on things, perhaps even more strongly than in spoken languages.

A different kind of hypothesis is that we humans, being human, prefer to talk about ourselves and things that are important to us, including other humans. We certainly talk about ourselves a lot; a recent study of English-speakers' conversations found that the word *I* was the most common word in their speech. A bias to mention ourselves first, and by extension other people who are important to us, could eventually grow into a general tendency to put any kind of actor early, so that nonhuman actors like dogs and avalanches would also tend to show up in sentences early, before whatever they're acting on.

Here we're seeing theories about how a language could get its properties from the nature of the human brain. The idea that the brain shapes language must have truth in it, but this particular explanation is still not fully satisfying to me. I'd like to know what property of our brains would cause us to lead with important concepts instead of saving the most important for last. Writers and orators often save the most important point for last rather than putting it first. Indeed, this save-the-best-for-last strategy is explicitly taught in composition classes as the optimal way to structure ideas. Since there are clear beliefs that saving the most important for last is a good strategy in some kinds of talk, we need to understand why, in the basic, everyday sentences throughout almost every language in the world, actors are mentioned early, before what's acted on.

I think that these worldwide talking patterns come from two components of talk planning: getting words from long-term memory and ordering them in a sentence. First, getting the words: Memory research shows that words referring to humans, like *girl* and *teacher*, are retrieved from memory faster than words referring to inanimate things like *chair* and *water*. Human words get an additional speed boost because we talk about humans a lot—easy words become more common and therefore come out of long-term memory even faster.

Now, putting the words in order. Remember just-in-time planning and easy-first? In chapter 3, we saw that these partners shape how we perform all kinds of actions, from talking to jazz improvisation to toothbrushing. Just-in-time planning quickly moves us from planning to actual talking, getting words off the sentence assembly line quickly and continuing to plan upcoming parts of the sentence while overt talking begins. Easy-first is the bias toward starting sentences with whatever easy words have come out of memory first.

These processes of planning our talk give us a very different idea about why actors come first in the basic sentences of almost all the world's languages. It's not that we *like* to put humans first or *decide* to put important ideas first. These tendencies aren't deliberate; they turn out that way because of the unconscious processes planning our talk, which is made more efficient by starting with the words that come from memory quickly.

A possible objection to this story is that maybe the easy-first bias somehow makes it easier for us to comprehend language. If so, maybe talkers arrange sentences the easy-first way to improve their audience's comprehension. And by extension, maybe languages in general change more to improve comprehension, not to make talking more efficient. I don't think so; easy-first and other good-enough talking strategies really do seem to be aimed at helping the talker. We see easy-first in actions that don't have an audience component, like toothbrushing, so we can be sure that efficient action planning really does benefit the person doing the action. And remember that some good-enough talking actually creates more work for our audience, such as when they have to figure out the meaning of ambiguous words. And finally, we know that talking is harder than comprehending, and making the harder task easier makes conversations flow better, benefiting everyone.

Having a Flexible Grammar Benefits Talkers

Although linguists have extensively documented and discussed the strong bias toward putting actors before whatever is acted on in basic sentences, you may wonder why they think that this basic order is so special. We have so many other kinds of sentences and word orders.

For example, if we see a dog and cat running around, we don't have to use the basic order and say, "The dog is chasing the cat." Instead, we could say, "The cat is being chased by the dog," which is a passive sentence. Look what we just did there: We've just said that English's basic order is actor, action, acted-on, but passive sentences flip that order and put the thing acted on (the cat) first and put the actor (the dog) last.

You may have heard from English teachers that passive sentences are evil and should be banished at all costs. Writing style guides concur—avoid passives! We'll examine that advice shortly, but my point right now is that English teachers and style guides throw shade on passives because we talkers use them fairly often. If we never said them, no one would be complaining that we use them too much. Talkers also veer off into using many other kinds of sentences that don't follow our basic English word order of actor, action, acted-on, even though the basic order could have conveyed our idea.

You may be thinking that I've just talked myself into a corner here. I began by saying that easy-first during talk planning gives us the actor-first bias in the world's language and that helps talking efficiency. But now I'm saying that we also produce passives and other nonbasic word orders. What's the point of having a basic, efficient option if people don't always use it?

Linguists suggest one answer: Having a variety of sentences lets us convey different nuances in meaning. On this view, talkers abandon the basic word order when some other word order can convey their ideas better. For example, the passive sentence *The cat is getting chased by the dog* draws extra attention to the cat's situation, in contrast to *The dog is chasing the cat,* which emphasizes the dog more.

Because passives and basic order sentences can convey different shades of meaning, blanket advice to avoid passives is a bad idea. Not only does strategic use of passives provide the variation in sentence structure that writing style guides also want, but passive sentences can be quite handy for emphasizing whatever's being acted on in the event: Wow, that poor cat's getting chased all around the yard by some dog, and nothing emphasizes the cat's situation better than a passive.

I like the fact that linguists' analyses can rebut the oversimplified advice to expel passives from our lives. Nonetheless, I'm chafing a bit about the cause and effect in their explanations here. Linguists suggest that we choose certain kinds of sentences in order to get our meaning across. Yes, we can choose words and word orders deliberately and carefully if we need to, but we've already seen that in most circumstances, talkers go with some good-enough option that makes talking easier. Therefore, I don't think it's plausible that we're typically pondering subtle shades of meaning and choosing passives to fit them. Instead, I think that passive sentences just naturally emerge from just-in-time planning and easy-first.

Remember that for planning to be efficient, we've got to be able to begin our sentences with whatever words come out of long-term memory early. That fact gives us an explanation for why languages have a variety of sentence types—the basic word order won't always work with what word comes out of memory first, and we need other options to let us be efficient talkers. For example, if you're discussing your toddler's first experience of a thunderstorm, you quite naturally might get your toddler's name from long-term memory faster than the word *thunder*. If you then take the efficient path of beginning with the easy, first-retrieved word, you get a passive sentence that begins with your child's name, such as *Maya got scared by the thunder last night*. It's not that you deliberately opted to emphasize Maya or decided to say a passive. It's just that, given her importance and familiarity to you, her name naturally comes from memory first, and the word ordering follows. This is another way that talking shapes languages: Having a variety of sentence types, including the basic word order, passives, and many others, serves our talk-planning efficiency because we can lead with the easy words and make a sentence that conforms to that order.

An unusual experiment at the University of Pennsylvania gives us some insight into the relationship between getting words out of long-term memory first and the kinds of sentences we say. In the study, research participants had to describe simple drawings that appeared on a computer screen. These drawings showed two figures and an action, like a dog chasing a cat. In a clever twist, the researchers used subliminal images to manipulate the talkers' attention to different parts of the drawing.

You may have heard of subliminal images in advertising, such as the words *Buy popcorn* appearing for a fraction of a second in an advertisement in a movie theater. *Subliminal* means without conscious awareness, and audience members are not typically aware of the brief *Buy popcorn* message on-screen. Nonetheless, the words affect their behavior—popcorn sales tend to increase after the subliminal words are shown, at about the same rate as when *Buy popcorn* appears on the screen long enough for everyone to notice the words.

Researchers used simple subliminal images in the talking study to direct the research participants' attention to different parts of the computer screen. Before each scene appeared on the screen, a small black box appeared for less than one-tenth of a second, too brief for anyone to become aware of it. But as with the *Buy popcorn* example, the subliminal box affected people's behavior. Their eyes tended to move toward where the box had been. Half the time, the box had appeared on the side of the screen showing the actor, such as the dog who's chasing the cat. The other half of the time the subliminal box appeared on the other side, which got people looking over to the cat. This means that the talkers' eyes and attention were directed to look at one image (dog or cat) first, and as a result, they got that word out of their long-term memory first. According to easy-first, talkers should tend to start their sentences with the first word they get out of memory. That's exactly what happened: If the rectangle was near the dog in the picture, talkers tended to begin their sentence with *The dog . . .* , saying something like "The dog is chasing the cat." When the rectangle was on the cat side of the screen, participants tended to describe the picture with the cat first, saying something like "The cat is getting chased by the dog." Efficient easy-first talk planning shapes languages, promoting a language having a variety of word order options.

The importance of having these options isn't limited to whole sentences; we can see that much smaller phrases have word-order options too. Remember the yellow-brick road in the movie *The Wizard of Oz*? The film was based on a book written some forty years earlier, where the famous road had a different name, used again and again as Dorothy and her friends traveled along it to find the wizard. In the book, it's called *the road of yellow brick*, and the phrase *yellow-brick road* never appears even once.

The screenwriters changed the words around for the movie, and today *yellow-brick road* sounds much better to us. That's partly because the movie is more familiar than the book, but that's not the whole story. Another reason *yellow-brick road* sounds better is that English is gradually changing to more consistently have adjectives and other description words before the thing being described. You probably can feel the difference: *yellow-brick road* sounds much more modern than *road of yellow brick*, which sounds stilted and old-fashioned.

This gradual change in word-order patterns probably has many forces shaping it, and I don't pretend to understand all of them. I do believe that however the trends go, English will continue to have word order options. *Road of yellow brick* might sound weird to us now, but there are still plenty of situations where we tend to put the description words at the end, as in *city of lights*, *secretary-general*, *road with too many stop signs*, and so on. That flexibility of word order is important for efficient talking, and so it will stick around in English and other languages.

French, meanwhile, generally likes to have adjectives after its nouns—in French, *the yellow-brick road* would instead be *la route de briques jaunes*—*the road of bricks yellow*. But as with English, French has exceptions to their noun-first order. A few dozen French adjectives typically come before their nouns, as in *la petite route* (the little road). Probably several forces are at work to create these exceptions, but the biggest one is talking practice: The most common adjectives, the ones that come out of memory the fastest, get to go first, before their noun. I think we can see an easy-first pattern in talk planning again shaping language here.

Language Change and Language Design

If languages evolve to get more fit over time, why should we wait for languages to get better gradually? In particular, if we know that talking is hard work and we are aided by language features that make talking more efficient, couldn't we rejigger our language, or invent a new language to accommodate what we now know? This might sound good in principle, but it's another thing to make it work in practice.

On the practical side, societies don't do well with radical change on the scale of wholesale language reconstruction. We cling to many imperfect inventions even if giving them up and switching to a new system would lead to more efficiency in the long run. Consider our QWERTY keyboard for typing and the imperial system of measurement used in the United States and a few other countries—quarts, inches, miles, and so on. Both of these inventions needlessly slow us down, but we're not ready to suffer the short-term trauma of abandoning them, even though more efficient alternatives exist, like the Dvorak keyboard and the metric system. And particularly for language, folks can become quite hostile to any proposed changes to their way of talking, a point that we'll explore in more detail in chapter 8.

A few invented languages do exist, but none to my knowledge have been explicitly designed to increase efficiency. There is, however, a fictional language aimed directly at talking efficiency, and I think that it's instructive to take a look at it.

The language is called Speedtalk, and it appears in a 1940s Robert Heinlein novella. I first read this mediocre tale in my youth, and I probably remember it today only because of the language angle. Heinlein tells the story of a secret group of high-IQ conspirators trying to revolutionize society to better suit geniuses like themselves. The novella's title, *Gulf*, refers to the intellectual gulf separating the geniuses and the everyday people they scorn. The super brains think that existing languages are outmoded and suitable only for non-geniuses, and so they invent Speedtalk. It was designed to optimize the efficiency of their communication and thinking, further supercharging their intellectual superiority. Gee, it's kind of like the "talking tunes your brain" theme in this book, but in a fictional authoritarian cult scheming to murder someone on the moon.

Gulf's plot is pretty thin, and so is the language. All of Speedtalk's efficiency came from making its two hundred thousand words really short, most only one syllable long. In the story, the geniuses think that this idea is life-changing, but they've got it way wrong. Speedtalk hits the trifecta of bad design—it would have been impossible for even a genius to learn, understand, or talk efficiently.

There aren't enough syllables in any language to have a unique syllable for each of two hundred thousand words, and so the Speedtalkers had to make words out of very similar-sounding syllables. They might have the short word *rog,* but also another similar-sounding word *rôg* (higher pitched *o* sound), *rrog*, and *rogg* (longer *r* or *g* sound), *rôgg, rrôg, rrôgg, rôôgg, rrôôg, rrôgg*, and on and on with many variants of *rog* and of other syllables, enough to make two hundred thousand very short words. Listeners wouldn't be able to tell many of the similar words apart, and they'd constantly need to ask for clarification. That in turn would make talking slower, because speakers would have to repeat themselves and enunciate carefully to make themselves understood—the opposite of efficient talking.

By focusing solely on word length to gain efficiency, the geniuses made an invention that was distinctly non-genius. It's certainly true that talkers do shorten some words, like converting *cryptocurrency* to *crypto*. Short words can be said more quickly, but notice that we talkers don't shorten everything. The people who pushed for saying *crypto* were those who were using the word frequently. In general, it's only the more frequently used words that get shortened. Speedtalk's problems show that shortening every word creates chaos.

The geniuses' second design error was focusing only on overt speaking. Planning our upcoming talking is a huge source of mental effort, but research on talk planning didn't exist when Heinlein was writing the story. We now know that having to work with many very similar words leads to more trouble finding a word and more errors while talking. Having thousands of very similar short words would leave the Speedtalkers navigating constant tongue twisters, again killing efficiency.

Heinlein's story and the Speedtalk idea are almost eighty years old now, and it's worth considering whether we could try language engineering again, given our greater knowledge about how languages work. We will see a few examples of requests to change our wording in the next chapter, but in general, I don't think wholesale reinvention is going to fly. "Everything is more complicated than we first assume," says geochemist Hope Jahren about her own field, and language patterns, their evolution and change, are also perfect examples of hugely complex systems that defy simple tinkering. Languages naturally become fitter over time, so improvements are already happening.

CHAPTER 8

By Your Talking You Shall Be Judged

Zora laughed. "You should hear my dad's freshmen. *I was like*," she said, pitching her voice high and across the country to the opposite coast, "*and then she was like, and then he was like, and I was like, oh, my God*. Repeat ad infinitum."

ZADIE SMITH, *ON BEAUTY*

THE FAMILY ON MY FATHER'S SIDE IS SCOTS ITALIAN. THIS blend began in a suburb of Boston when Eleanor Sarno, teenage daughter of Italian immigrants, started to work for a large family of MacDonalds, who hailed from Scotland by way of Canada. The matriarch of the household was my great-grandmother, the family's first Mary Ellen MacDonald. Young Eleanor cooked and cleaned and lived in fear of the formidable Mary Ellen, an alcoholic who routinely went off about the inferiority of Italians.

Others in the family appear to have been more friendly, and Eleanor got pregnant and married one of Mary Ellen's sons. She and her husband inherited Mary Ellen's house and raised five children there. One of them was my father.

Eleanor's marriage gave her a middle-class life and an escape from poverty and her abusive father. Given that welcome transformation and Mary Ellen's rants about low-life Italians, it's not surprising that Eleanor wanted to leave much of her heritage behind and assimilate to American culture. Nana, as we grandkids called her, was bilingual but exclusively spoke English. She had an intense aversion to garlic and cooked only a narrow subset of the delicious Italian dishes she grew up eating. Instead, she extolled the virtues of Yankee cooking and prepared items such as New England boiled dinner, which tasted about like it sounds.

A really striking part of Nana's assimilation was that she changed the way she talked. She never graduated from high school, but she spoke what she felt was more proper English than the speech from her childhood. Colloquial phrases like *I says*, which could be heard in her sisters' speech, were banished from Nana's repertoire. And she forced far more *r* sounds into her speech than could be heard in the Boston accents around her. She called herself *Eleanor*, with the *r* intact, while everyone else said *Elen-ah*. Her accent was still identifiably New England, but with some of the immigrant working-class features sanded down.

Adopting those *r*'s was a notable accomplishment, because changing even part of an accent can be very hard to do. Some folks who want to leave behind their family's way of talking seek out lessons or tutoring because they can't seem to escape their accent on their own. My father also deliberately tried to lose some of his Boston accent but didn't have Nana's success with the *r* sound. He resorted to steering clear of problematic words like *park* and *car*, which are famous for being pronounced *pahk* and *cah* in Boston. This avoidance forced him into awkward workarounds such as—I kid you not—*sports automobile.*

Nana succeeded while her son failed in accent modification for a couple of reasons. She lived with her Scots Canadian in-laws, who probably had distinctly non-Boston accents, with plenty of *r* sounds. And Nana was bilingual in Italian and English, while my father spoke only English. Bilinguals, by virtue of their experience navigating different pronunciations, words, and grammars across their two languages, often become quite adept at developing new language skills.

Nana was driven to change her talking because she wanted to signal that she was someone who belonged in the MacDonald family. She understood that the way people talk reveals their tribe—their nation and region of origin, race, social class. I'm using the word *tribe* quite deliberately here, because it conveys a group that's cohesive and identifiable, with deep interconnections, distinct from other groups.

Our talk broadcasts our tribe identity rather like wearing a sports jersey identifies an allegiance to a sports team. If you see someone walking down the street wearing team merch, you can immediately infer something about them even if you've never met them. Similarly, our talk reveals our tribe immediately, before anyone gets to know us.

Of course, wearing a sports jersey doesn't mean that you have no other identity, and similarly, you're more than simply a member of your tribe. You're a unique individual. In chapter 9, we'll see how features of your own unique style of talking reveal aspects of your individual identity. The patterns linking your own unique talk to you are very subtle. That's an important difference between individual and group talking patterns: An individual's talking characteristics require complex computer programs to identify, but the talking patterns we share with our tribe announce themselves to whoever encounters them.

Being branded as a member of a group first and individual second can have real consequences for how we're treated. On the plus side, when you meet new folks from your home area, your talking will signal that you're one of them. This immediate connection will tend to promote getting to know each other, offers of mutual aid and friendship, and other benefits. On the more negative side, your talk can identify you as an outsider, potentially leading to misunderstanding, distrust, and even hate and violence. It's important to understand how ways of talking can range from beneficial to dangerous, depending on the situation.

In this chapter, we'll see why communities end up with identifiably different styles of talking. We'll explore the very common opinion that someone's own way of talking is better than other ways that they encounter. We'll see when this prejudice begins and how it affects our lives.

Why Talk Signals Tribe

The parts of our talking that broadcast our group identity are precisely the ones that are hard for us to change. We can change our clothes, our jobs, and where we live, but the way we pronounce our *r*'s probably isn't going to budge. As a result, these deep-seated talking qualities tend to become the clearest signals of our original tribe.

The first tribe-signaling component is accent, the specific way we pronounce words. The word *pronounce* is usually associated with speech, but there are also accents in sign languages. Sign language accents affect how words are articulated with the hands, body, and face. As with accents in speech, sign language accents crop up in geographical regions and can vary by class and race.

Accent is part of a larger component of our tribe's style of talking, called dialect. Dialect includes not just accent but can also include tribe-specific word orders, intonations, politeness traditions, and word choices. For example, there are three regional names for carbonated soft drinks in the United States: *soda*, *pop*, or *Coke*, and you probably use whichever term you grew up with. I am for *soda* all the way, but I've lived in both *pop* and *Coke* dialect regions. Saying the "wrong" term can brand you as an outsider real quick.

Accents and the other components of dialects are learned from our conversations and interactions with other talkers, beginning in childhood. The childhood period is so important for establishing a dialect because children are both the most agile language learners and the most local talkers. We adults encounter many dialects and accents through travel, work, and media, but kids' experiences are much more restricted; they primarily talk to family and other local kids. The "other kids" is the important point here. Children are actively learning and reproducing what other kids are saying, and a child's dialect will typically reflect the talk of the kids around them more than their own family's way of talking. Kids learn their dialect from their peers for all the reasons why kids learn so many other things from their peers—they're interacting socially, talking to each other, maneuvering for position, trying to find their place in their social world. Adapting to their social milieu, including in dialect, is part of children's social development.

Once this local style is entrenched, it's very hard to change, in the same way that it's very difficult to learn a new language later in life.

When these dialect patterns have solidified, they make the talker's tribe crystal clear.

Judging the Out-Group

We humans often view folks outside of our own group with suspicion and even hate. The presence of almost any difference seems sufficient reason for wariness, including variation in race, religion, language, customs, or location. Distrust of out-groups is a complex enough topic that it could fill another book, and we won't try to delve into its many facets here. What I do want to explore here is how talking affects, and is affected by, all this in-group versus out-group partitioning and distrust.

I especially want to get at the fact that many folks dislike other groups' ways of talking. This dislike of others' talk might simply stem from talking's function as tribe signal—someone distrusts members of an out-group and naturally extends this prejudice to everything associated with them—their talking, cuisine, clothing, music, whatever. In this situation, hating an out-group's talking is sort of like blaming the messenger that signals the out-group status. Alternatively, some people may hate unfamiliar talking all on its own, above and beyond any ways in which it signals out-groups. We're going to explore these possibilities.

Whatever the nature of this prejudice against other ways of talking, it starts really, really young. Five-month-old infants, pre-talkers not even old enough to sit up on their own, have clear preferences to listen to people who are speaking with familiar accents than to those with unfamiliar ones. And when asked to choose children as friends, young children pick the ones who speak in familiar accents and shun the ones with unfamiliar accents. When race and accent are pitted against each other in these friend-choosing experiments, accent wins over skin color—white five-year-olds readily choose Black children speaking with a familiar accent for potential friends over white children speaking with an unfamiliar accent. These childhood biases favoring the familiar accent don't go away with broadening experience: Young children from multiethnic communities who encounter many accents at school still choose children speaking in their own immediate group's accent as potential friends over children speaking with a different but very familiar accent.

These accent biases just keep coming as we grow up. There is so much research on talking-based discrimination and hatred that new academic journals have been started to keep up with the demand. When I was sorting through this literature, search engines offered helpful suggestions for the many types of discrimination I might be looking for: Did I want research on discrimination against immigrants' accents? Discrimination against white Southern US accents? Juries disbelieving witnesses with low-status accents? Might I want the research on accent discrimination in the nursing profession, or in dozens of other professions? Accent discrimination in the United States or other countries? Talking-based discrimination is abundantly documented in all of these, and more. It affects people's housing, schooling, hiring, promotion, medical care, economic prospects, and on and on.

Much of this research owes its start to the work of the linguist John Baugh, a professor at Washington University in St. Louis. In the 1980s, Baugh accepted an invitation to spend a year at Stanford's Center for Advanced Study in the Behavioral Sciences. He began calling rental agencies near Stanford to find a place where he and his family could live for a year. Four appointments were quickly arranged, but when he showed up at each of them, he was told, sorry, the place was already rented. Baugh strongly suspected that he was welcomed over the phone but turned away in person because his talk did not signal the same tribe that his appearance conveyed. Baugh is Black, but his accent and dialect don't identify him that way. A lawyer told him he should sue, but Baugh instead saw a research opportunity.

As a child in Los Angeles, Baugh noticed that his mother talked very differently on the phone to white people and Black people. She was able to code-switch, which is what linguists call the ability to shift from one language or dialect to another in different social situations. Baugh's mother could adjust her dialect to align either with the Black English of her community or with what's often called General American English, spoken by many white people in the United States. Baugh typically speaks this General American English dialect, but he can code-switch to Black English. And his keen ear for talking styles also led him to recognize and imitate the nuances of the Latino English he encountered growing up in LA. He decided to put these code-switching skills to work in a study of talking and discrimination.

The US Fair Housing Act, passed in 1968, made it illegal to discriminate in rentals or sales of housing. Initially, rental agencies that were sued for racial discrimination argued that they couldn't tell anyone's race over the phone, so they weren't discriminating if they turned people away who called for appointments. They were essentially saying that someone's talk doesn't signal their tribe. That's a laughable position now, but at the time John Baugh began his research, there was little proof. Baugh wanted to collect some solid evidence.

He set up an experimental protocol for himself in which he called rental agencies that had advertised rentals in several different California communities. He called each agency three times, several hours apart. Baugh opened each phone call with the same words: "Hello, I'm calling about the apartment you have advertised in the paper." Sticking to these exact words every time kept the dialect constant in each phone call. What varied was the accent he was using.

In each call, Baugh spoke with either a Black English, Latino English, or General American English accent. He recorded whether the person on the phone said the apartment was still available or already rented. After tabulating the data from hundreds of phone calls, Baugh found that in every single one of the communities he called, people at rental agencies said the apartment was available for viewing far more often in response to his General American English accent than to his Black or Latino accents.

Baugh's results showed not only that listeners could make rapid inferences about someone's tribe on the basis of their talking on the phone, but also that many of them discriminated against people with Black and Latino accents, without having met the person. Since Baugh's initial study, he and many other researchers have documented widespread accent-based discrimination in phone calls related to housing, employment, obtaining insurance, and getting medical care.

What's the Matter with Other People's Talking?

The piles of studies like Baugh's reveal the staggering range of situations in which people pay the price of talking differently than someone else. What's missing is a firm understanding of why the talking style of an out-group is so reviled. An obvious interpretation, as we discussed earlier in this chapter, is that talking is simply a signifier of someone's tribe and out-group status. It's undeniable that prejudice and discrimination against individuals and their groups exist; talking may only be the messenger.

The talk-as-messenger idea is likely real, but I want to explore an additional angle, that there is also prejudice against another tribe's talking itself, layered on top of prejudice against individuals or out-groups. If that's true, then differences in talking could not only identify out-groups but also amplify the prejudice against them. And dislike of someone's talking style could crop up even among people within the same tribe.

We humans have great affection for own traditions, rules, and laws, and that's often a good thing. Our sports would be no fun without rules of the game, our economy couldn't function if our currency had no rules for its value, we need drivers to stop at stop signs and not run pedestrians off the road, and so on. Sure, people rail against bureaucracy and red tape sometimes, but these folks still wouldn't want monetary chaos or cars running people over. Rules, traditions, and laws give a necessary structure to our lives.

Some rules and traditions are rooted in basic moral beliefs, but others are more arbitrary accidents of history. Many societies signal respect for the deceased by wearing certain clothing at funerals, but choices of attire vary widely across the globe. In most Western cultures, wearing black is traditional, while in many East Asian and South Asian cultures, mourners wear white. In still other cultures, funeral attendees traditionally wear yellow, red, or another color. In each case, the color choice is a culturally specific communicative signal of mourning and respect, enshrined by centuries of rituals.

Despite the importance of particular mourning colors within different cultures, most of us can easily believe that mourners wearing white in one culture are no more or less respectful than mourners in black in another culture. And looking across all those cultures, we can see that no color choice is inherently better than any other in communicating the wearer's message of mourning and respect.

Given these insights about funeral attire, I want to consider whether we can be similarly evenhanded with another form of cultural diversity, dialects. Are all the dialects of a language to be considered societal traditions on par with funeral attire, deeply important within a culture but with no one alternative inherently better at communicating than another?

Let's consider this question via English dialects. English is often described as having about 150 major dialects around the world, with about thirty dialects in the United States alone. Each of these major dialects has many subdivisions, because dialects can emerge within region, race, class, and any other force that creates somewhat isolated local groups reinforcing each other's local talking styles.

With all these dialects and subdialects, there are abundant opportunities to ask whether they are all equally good as a means of communication. Linguists have extensively studied this question, and the results are extremely clear: Yes, all dialects are equally good for turning our internal messages into external talk. Dialects differ widely in their social status, but as linguistic systems, they are all equally effective for talking within their own culture.

This view is not widely shared outside of the linguistics community. The linguist Lisa Davidson of New York University says that prejudice about the way someone talks is one of the last "acceptable" prejudices in our society, able to be freely expressed without embarrassment or pushback. Surveys about talking prejudice bear her out. Around 60 to 80 percent of Americans readily volunteer that they have negative opinions about people who speak what they view as lower-status dialects of English, such as Black English and Southern American English. These same people might reject prejudice (or hesitate to express it) against race, age, sexual orientation, weight, or religion, but someone's talking? Yeah, we don't like it, and we're not afraid to say so.

Some of this talking prejudice may stem from having trouble understanding certain dialects or associating "better" dialects with amount of education and social status. And some prejudice toward a dialect comes from prejudice toward ones who speak it, such as low esteem for Black English stemming from racial prejudice. But I think we can make a case that many people really hate unfamiliar talking itself, beyond any dislike of the ones who produce it.

This talking-specific prejudice stems in part from our love of rules and traditions. Languages are complex systems with many moving parts, and we've all been taught in school, and perhaps by other members of our tribe, that there is one right way to put those parts together. It's astounding that the same elementary school classrooms that abundantly celebrate the diversity of customs and holidays across cultures teach kids that there is exactly one right way to talk. If folks have embraced the "one right way" view of talking, then a natural response to another system of talking is to think of it as wrong. Talking doesn't just signal tribe here, it also signals that this talker is ignorant of the rules, is talking incorrectly, is violating the norms and traditions of our society. That's what I mean by prejudice against a form of talking itself.

Evidence for this idea comes from observing how folks start to judge the talk of those within their own tribe. Our language and our style of talking are always changing. Young people, who have had less practice with their local dialect, are the ones who tend to be more open to change when they hear their peers innovating some new expression or intonation. Older folks and anyone who's deeply into rule following often push back.

This cycle of change and resistance is as old as talking itself, or at least as old as written records can reveal. Some 2,500 years ago, in what we now call India and Pakistan, ancient linguists worked to understand the grammar of the Sanskrit language and write it down for posterity. Sanskrit was the sacred language of important religious texts and poetry, and the scholars were alarmed to hear the everyday talk around them drifting away from the classic Sanskrit forms. These early linguists believed that writing down the grammar would prevent the language from being ruined by the young folks who were screwing it up. They created incredibly insightful analyses of Sanskrit that are important to linguistic scholarship to this day. The ancient grammarians did not, however, stop the change of language. No one can.

Pretty much any aspect of language can undergo some noticeable change within the span of a talker's lifetime. Some change is simply from necessity of inventions; the words *smoothie* and *Covid* came into wide use because we had to talk about these new aspects of our lives. Other words, like *swole* and *OG*, don't refer to new concepts; they're new words adopted by young people to supplement or replace older ones. Replacing an existing word with a new one leaves older folks behind, parked somewhere between clueless and peevish. That may be the point; young people want to innovate and abandon old traditions, poke a little fun at the users of old-timey phrases that went out with VCRs.

One talking trend that really irks older English speakers is the decline of saying *you're welcome* after someone says *thank you.* All over the world where English is spoken, younger people's responses to *thank you* are now commonly *no problem* or *no worries.* Those who are firmly on Team *You're Welcome* appear remarkably upset that this time-honored expression has been jettisoned, seemingly for no reason. Even language researchers, who surely know that language continually changes, have complained to me about it.

Some irritated folks interpret *no worries* and *no problem* literally, complaining that they weren't actually worried when someone says *no worries* to them. They may even say *no problem* themselves in a different situation, when someone apologizes. These folks likely find it really idiotic that others appear to mix up thanking and apologizing. I think it's helpful to remember that thanking routines in English and other languages are more politeness rituals than informative talking. After all, *thanks* and *thank you* aren't exactly grammatical English sentences outside of the context of this ritualized exchange; they're shorthand for something like *I acknowledge your kindness.* And responses to thanks are also ritualized, without much literal meaning: *you're welcome, it's nothing, not at all, think no more about it, my pleasure, no problem,* and *no worries.* These are all ritualized signals of acknowledgment.

The irritated among us also fret that these modern phrases seem too informal and less polite than good old *you're welcome.* But a shift in the words signaling politeness is not the same thing as the decline of politeness. *Hello* was once considered a rude and overly casual greeting compared to the then-polite *Good morning* or *How do you do?* Complaints about *hello* back in the day did not prevent it from becoming a common and polite greeting ritual. Indeed, *hello is* now viewed as more polite than alternatives like *hi*, *hey*, and *what's up*. Similarly, *no problem* and *no worries* are on a path toward becoming polite components of our thanking rituals. If you're irritated by *no worries*, perhaps you can learn to be at peace with the fact that change is inevitable, and the new and unfamiliar need not be bad. Same difference, no problem, move on.

Two other talking changes that get pushback aren't shifts in wording; instead, they are intonation patterns that are increasingly common in young English speakers—vocal fry and uptalk. Before we get into how they work, let's just note that a shocking amount of bias and judgment have been leveled against people who use these intonations. If you search for uptalk and vocal fry information on the internet, you'll find not only endless commentary about how stupid people sound for talking this way but also instructional videos to help people banish these supposedly evil patterns from their speech. The level of hostility and alarm they elicit seems more appropriate for combating serious social problems like underage smoking or drug abuse, but here we're dealing with the supposed evils of pitch changes when speaking. Rilly?

Hysteria over something as innocuous as speech intonation shows us how talking itself is policed and judged, and how inflexible some folks can be in the face of change. Published anti–vocal fry and anti-uptalk editorials actually reinforce the triviality of the issue. In contrast to massively complex societal problems like drug abuse, both uptalk and vocal fry are seen as simple youthful rebellion, to be stamped out simply by telling the rebels to get their vocal cords back in line. The etiquette columnist Miss Manners once received a letter asking for advice on how to tell a young woman to get rid of her vocal fry. The letter writer's clear assumption was that of course this way of talking is horrible, and the only thing needed to fix it was for Miss Manners to describe how best to tell someone to stop already.

Once we understand the nature of vocal fry and uptalk and how they're being used, it's clear that these pitch changes are not just affectations or youthful acting out. Like so many talking changes that we've seen in this book, they are clever adaptations to the fact that talking is hard work. Uptalk and vocal fry are helpful for the speakers who use them.

Vocal fry, or creaky voice, is caused by a trailing off of airflow while speaking so that the voice takes on a low-pitched, creaky sound. Many languages use vocal fry in different conversational situations. In English, it's most often heard toward the end of what someone is saying. In previous decades, vocal fry was commonly used by men in Great Britain who spoke the highest-status British English dialect, and no one judged it as annoying. Since roughly the year 2000, vocal fry has appeared in the speech of mostly young English speakers of all social classes in the United States and elsewhere, and suddenly there are endless complaints about it.

The bias against vocal fry has a gender twist to it. Quite a lot of research on vocal fry suggests that young men and young women have vocal fry in their speech in approximately equal amounts. However, those who complain about it assume that vocal fry is primarily a young woman's affectation. As a result, young women are the ones who are being judged for it, with descriptions of women's speech with vocal fry including "vulgar," "repulsive," "mindless," and "really annoying."

In controlled studies, listeners are equally good at detecting vocal fry in men and women when they're asked to listen for it. However, in everyday conversations outside of a laboratory, it's possible that the pitch drop of vocal fry tends to be more noticeable in women's voices than in men's voices. We don't know if that's true, but we do know that vocal fry is judged as more irritating in women's voices than in men's. This pattern fits nicely into a long-standing tradition of policing women's speech more harshly than men's. The linguist Alexandra D'Arcy, who studies changes in English over time, has argued that for many centuries, women have been blamed for essentially every change in English that old folks have complained about, even though both men and women use the new way of speaking.

Why would young women embrace vocal fry when it exposes them to rabid judging by older and potentially more powerful people? Because it's useful. There are two good theories about how it helps their talking. Because vocal fry lowers the pitch of someone's voice, it can make the speaker sound older and more authoritative. As young women tend to have higher-pitched voices than men, vocal fry may be more useful for them. Another idea is that the low pitch coming from vocal fry at the end of utterances can be a signal that someone is ending their speaking turn. Speakers of English already typically drop their pitch at the end of many kinds of sentences, and vocal fry would make that drop more obvious. These ideas could both be true, and it would not be surprising if the rise of vocal fry has several benefits behind it.

Uptalk goes in the opposite direction, with rising pitch at the end of certain phrases. It's similar to the intonation of a question in English, but it's used in statements. Uptalk is sometimes associated with speakers in California, but it appears throughout the United States and in many English-speaking countries. It has even been observed in Spanish, led by bilingual speakers who have uptalk in their spoken English and carry it over to the other language they speak.

And as with vocal fry, uptalk is widely perceived as an affectation in young women, but studies show it is widely used by both young women and young men. Indeed, it's been around long enough that I can hear it in plenty of middle-aged speakers too.

Like vocal fry, uptalk is helpful for the speakers who use it. It tends to appear at the ends of phrases that are not the end of someone's full conversational turn. For example, if an uptalk user were saying, "Yesterday we went to the store, and we ran into Jonah," the rising pitch of uptalk would be on the word *store.* The final word, *Jonah,* would get a falling pitch and maybe vocal fry, signaling the end of the conversational turn. Having an uptalk pitch rise in the middle makes it clear that the speaker isn't done speaking. Perhaps uptalk is especially useful for young women holding the conversational floor, as they tend to be interrupted more often than men.

We saw in chapter 3 that turn taking in conversations can be complicated for listeners as well as talkers. Now we're seeing that pitch changes, which are easy for talkers to implement on the fly, are helping to create efficient turn-taking signals in conversations. The uptalk in the middle of a sentence and low-pitched vocal fry at the end create prominent signals for the listener about where the speaker is in their talking turn, helping the listener predict the end of the turn and plan what to say. Vocal fry and uptalk do this helpful work even if listeners are not aware of their functions, and even if those listeners think uptalk and vocal fry sound stupid. Thank you, uptalk and vocal fry, for improving the efficiency of our conversations! No worries, they say.

No discussion of judging talking would be complete without examining the word *like. Like* has many different uses, some traditional dating back centuries and some newer, controversial ones. Everyone who speaks English probably uses *like* in many traditional ways, such as in *I like ice cream* and *It was like a dream*. Other uses of *like* are associated with younger speakers and are subject to quite a lot of judging.

The kinds of *like* that many folks hate include the *like* in *I was, like, totally weirded out*, or *They said it was, like, one hundred degrees.* More examples are in the quote that starts this chapter, where a young woman, Zora, is ridiculing the talk of slightly younger people, college freshmen. Some versions of these *like*s can be found in English at least as far back as the 1800s, but they've become much more common lately. A study of college students' speech found that *like* was the fourth-most-common word in their conversations.

Why is this kind of *like* invading the speech of the younger generation? It's the same basic motivation that's behind uptalk and vocal fry: *like* is doing something useful for them, helping to manage the difficult job of talk planning. In chapter 3, we saw that just-in-time talk planning encouraged beginning to talk early when just a little bit of the talk plan was ready, with planning of later parts when talking is underway. This method can be really efficient, but sometimes the supply chain breaks down, and we run out of planned words. Then we have to pause and scramble to get the next part ready to say. Rather than grinding completely to a halt, speakers often produce a filled pause—some little bit of sound that extends planning time while also signaling that the speaker isn't done talking. Common filled pauses in English include *uh* and *um*, and now *like* can be a filled pause too, as in *I think that these ideas are like* . . . Younger speakers have embraced this new role for *like* in English, and it is helping them do the more efficient just-in-time planning of their talk.

Since many people still think of these modern speaking styles as annoying rather than effective talking strategies, young users of uptalk, vocal fry, and *like* might consider what to do when interacting with an older person in authority, such as when applying for a job. No matter what dialect they use, every job applicant has work to do before an interview, including learning about the company and the potential job, developing thoughtful answers to likely questions, and practicing those answers to make them fluent. That preparation will naturally reduce the number of *likes*, *ums*, and other filled pauses in speech, because practicing an answer will mean that fewer filled pauses are needed for extra planning time during the interview. Practice may also naturally reduce some vocal fry and uptalk, because these signals of turn taking are less necessary in a fluent, practiced answer.

While some of these turn-taking and talk-planning features may naturally recede with interview preparation, I don't recommend following advice on the internet on how to eliminate *like*, vocal fry, and uptalk from speech. Most of the suggested techniques involve paying attention to words, breathing, and pitch, and this stuff is the last thing someone should be focusing on in an interview. Or in any kind of conversation. The magic of talking is that it works without conscious awareness of what's going on under the hood, driven by the ideas we want to convey. Turning our attention away from our message to focus on the mechanics of talking is a recipe for losing track of our thoughts and sounding the opposite of competent and fluent.

If you're a *like*, vocal fry, and uptalk hater, perhaps this discussion of their benefits will modulate your stance. But maybe not. Because our talk is such a strong marker of tribe and rule following, younger talkers' inventions of new traditions can make older folks feel like the talk is shifting out from under them. In some respects, that is exactly what's happening—young people are going out, leaving the nest, flouting old rules, claiming their own place in the world, including the world of talk. That's as it should be, and I encourage the haters to let it go.

But before young people get too smug here, they should remember that in a few decades, they too will feel left out and will be judging the newer generations' newfangled habits and talk. The Sanskrit grammarians resented language change but couldn't stop it thousands of years ago, and since then, the march of change has played out in the same way. Each new generation thinks that they'll be more enlightened, and maybe they are about some things, but everyone seems to hate changes in talk.

When *We* Decide to Change Our Talk

The changes that emerge in talking over generations are not typically deliberate, but there are exceptions, when some group decides we should change what we say. For example, the suggestion to use the title *Ms.* as a marriage-neutral substitute for the *Miss/Mrs.* titles for women has shown up repeatedly over several centuries. It made an initial appearance beginning in the seventeenth century but didn't catch on then. Several men again proposed using *Ms.* in the early 1900s and suggested that it would avoid the embarrassment of not knowing whether the woman they were addressing was married or not. *Ms.* didn't get far then either, but it finally did gain traction in the 1960s, when its use was framed in feminist terms: that it freed women from having to be categorized by their marital status. *Ms.* became the name of a magazine around that time, greatly increasing its visibility. With that visibility came scoffing and judgments launched against it.

Now, fifty years later, *Ms.* is commonplace and unremarkable, an example of how deliberate changes can be initially controversial yet grow mainstream over time. It's useful to keep the *Ms.* example in mind when considering how newer suggested changes are faring, including other gender-related ones. These include the title *Mx.*, often pronounced "Mix," as a gender-neutral replacement for *Mr.* and *Ms.*, appropriate for addressing anyone. There are also gender-neutral pronouns, replacing *he/she, him/her, his/hers* with something without a gender marking, typically *they/them/their.* Like *Ms.*, gender-neutral pronouns have been around for centuries, but rapid changes are currently underway. One in five Americans now report that they've encountered someone who uses gender-neutral pronouns.

Whereas encountering new expressions like *No worries* primarily affects the audience of someone who says it, suggestions for *Ms.*, *Mx.*, and gender-neutral pronouns are aimed at the talker—it's a request that the talker start using different words. Given that talking is harder than comprehending, and given the dislike of changes in talking we've seen, it's not surprising that there is resistance in some circles to using new pronouns, titles, names, and so on.

When someone requests the use of their pronouns, some people adapt relatively easily, and others find it incredibly difficult to discard old pronoun habits previously associated with the person making the request. As we've seen, talking is strongly driven by the habits of word use, and our reliance on common words for someone are being upended here. The folks who are most likely to make pronoun mistakes are the ones who have known this person longest and have had the most talking practice with their previous pronouns: old friends, parents, siblings, other family members. Young people have had less practice with an earlier set of pronouns, and they tend to make fewer mistakes. At least in part because they are adapting well, younger talkers tend to consider pronoun changes to be less momentous than older folks do.

The most common recommendation for learning to use gender-neutral pronouns or other pronouns is to practice saying the pronouns the person uses, thinking about that person and practicing referring to them with their pronouns. This is excellent advice, and indeed practice is essential for change in talking to happen. But even with this practice, change is not necessarily rapid, and the number of mistakes can be quite appalling. I think it's useful to consider why the pronoun change is particularly difficult, even more difficult than a change in title or name. The reasons come from how talking works.

Pronouns are part of a system of ways to refer to somebody—*I, me, my, mine, you, your, yours, they, she, he*, etc. Notice how a request to use different pronouns typically affects only part of this system. A person who uses *they/them* pronouns still uses *I, me, my*, and *mine* when talking about themselves, and you still address them as *you*, and you also still say *your, yours* as needed. You'll still say *we* when talking about the two of you together, and *they, them, their*, when talking about them in a group with other people. It's only one corner of the system that changes—*they, them, theirs*, to refer to this person individually, and only when talking about them to someone else.

I can't prove it, but everything I know about talking suggests that using new pronouns for someone is hard in part because only one part of the pronoun system is changing. We can throw a whole referring system out and start over fairly easily, such as when someone changes their name, or when we retired *Miss and Mrs.* in favor of *Ms.* But altering only a part of an integrated system is more difficult. In the pronoun case, the unchanged parts of the system (*you, we,* etc.) keep reinforcing the other original settings we have for this person. Mutually reinforcing systems like this are usually great for efficient talking, but that same benefit becomes a burden when trying to change only a portion of the system.

The closest analogy I can think of to this situation comes from English verbs, which also form a system. With the verb *walk*, for example, we have *to walk, is/are/was/were walking, will walk, I/you/we/they walk, he/she/it walks, walked*, and *has/have/had walked*. Most English verbs slot perfectly into this system, but a few dozen of them deviate in only a small corner of it. The verb *run*, for example, has most of the same pattern as *walk—to run, is/are/was/were running, will run, I/you/we/they run, he/she/it runs*, but *run* is different in a few parts of the system: There's no *runned* equivalent of *walked* and instead we have *ran* and *has/have/had run*. This partial deviation in the verb system, where most of *run* goes through its paces like all the other verbs but has a few exceptions, is incredibly difficult for children to learn, and also hugely difficult for adults learning English as a second language. The networks of related forms are so strong that it can take literally years of practice talking to say these exceptions correctly. I remember that one of my kids still occasionally said *bringed* instead of *brought* at seven years old. That long learning path is not at all unusual.

The exceptional level of difficulty of partial deviations to a system helps to explain why changing pronouns can be so prone to errors, even among those who are truly trying hard. Perhaps understanding the level of difficulty here affords some grace concerning misgendering errors—perhaps less of *Aunt Sue doesn't respect me and my pronouns* and more understanding that Aunt Sue is succumbing to the natural difficulty of changing part of a talking system after decades of talking practice. But Aunt Sue isn't off the hook, not by any means. She and anyone asked to use gender-neutral pronouns for someone must recognize the importance of the request and the really serious effort that's going to be required to honor it. I've included some suggestions for practice in the endnotes.

Outside of pronouns and *Ms.*, talkers are also frequently asked to change how they refer to groups, institutions, and policies. Some examples include the rejection of racial slurs, the replacement of the term *mental retardation* with *intellectual disability*, replacing *homeless* with *unhoused*, *killer whale* with *orca*, and many more. Talkers now need to sort through options like African American versus Black, Latina/Latino/Latinx, people of color, BIPOC, and so on. There are now style guides to help journalists, spokespeople, company employees, and others avoid problematic language. There is also plenty of pushback against these changes.

Asking talkers to change their terminology is complicated. Many people likely find some new wording suggestions extremely worthwhile and others ridiculous. There won't be universal agreement on which changes are in the worthwhile category versus the ridiculous one. That's quite natural, because talking changes are routinely controversial. It's probably better to take a longer view of these controversies, remembering that the use of *hello* and *Ms.* were scorned initially but after many decades, they became commonplace. Other wording suggestions might remain controversial for years but might become more mainstream some decades down the road.

We should brave the controversy of new terminology because words have power, and in particular their power to denigrate others is real. But we should also keep in mind what we've seen in this chapter: talking, including word choice, is an in-group–out-group signaling system. If one group starts using new words even for benevolent reasons, this change naturally reinforces the in-group of the new-word users and excludes those who aren't hip to the new terms. And once words take on this function of sorting people into in-groups and out-groups, there's some danger that the sorting function can grow too large, turning into gatekeeping and diminishing the initial value of the change: We can't make rapid progress solving the horrific problem of people chronically living on the streets, but one group might criticize others who don't know enough to call such people *unhoused* or *houseless* rather than *homeless*. Words can be fists, as the novelist Sigrid Nunez observes in *The Friend*. Better not to also make them dividers between people otherwise working toward the same goals.

IN THIS CHAPTER, WE'VE REPEATEDLY SEEN THAT TALKING IS full of tribe-level markers that have the power to unite and divide us. What we haven't talked about is the ways that our talk can identify individuals within a tribe. Those individual markers exist; they just get swamped by the glaringly obvious tribe-level patterns of accent, word choices, and more. Researchers who study talking patterns can nonetheless find individual talking fingerprints that reveal quite a bit about the talker. Do you want to be tracked by the way you talk? You'll see more about what that means in the next chapter.

CHAPTER 9

The Science of Talk Analysis

Black words on a white page are the soul laid bare.

GUY DE MAUPASSANT

A FEW YEARS AGO, I RAN INTO SOMEONE I HADN'T SEEN FOR a long time.

He looked down at my feet and said, "Wow, what happened to you?"

"Well," I answered, "I just had foot surgery, and I have to wear this big boot for about six—"

That's as far as I got, because this guy grabbed the conversational reins and galloped off on his own talking path.

"Oh!" he said, getting really animated. "The sister of someone I know had foot surgery, and it was just a total horror show. She . . ." He launched into an extended tale of podiatry disasters experienced by this woman, someone neither of us had ever met.

Even though talking is hard work, Podiatry Guy, as I now think of him, clearly has a need to be doing a lot of it. Why is that? The question is quite interesting to me, in part because I'm absolutely not one of these must-talk people. Yes, I study talking. Yes, I love writing, which is a form of nonstop talking. And I enjoy lecturing and presenting my research to an audience.

But extended holding the floor in face-to-face conversation, regaling people with story after story? That's not me. Someone with higher talking needs might have steered that podiatry conversation back to their own foot, but I didn't go there. It didn't seem worth the effort to maintain the attention of someone who wanted to talk more than to find out what I had to say.

Podiatry Guy isn't alone in his burning need to talk. Experts in overtalking estimate that as many as one in five US adults are overtalkers, those who talk significantly more than would be typical for the social situation they're in. In light of their prevalence, it's not surprising that the rest of us have invented phrases to describe overtalkers and the experience of being dragged along in their conversational ride. Besides *overtalker*, we've got *oversharer*, *talkaholic*, *motormouth*, *talked nonstop*, *running at the mouth*, and *you can't get a word in edgewise.*

A more technical name for overtalkers is *compulsive communicators*, which captures the idea that these people can't easily change their overtalking behavior. Talking apparently fills a need for these people that is not easily abandoned. One overtalker told me that they had been seeing a therapist for decades about overtalking and related issues, yet they knew that they still talk too much.

If overtalking is associated with certain personality traits, can we gain insight into Podiatry Guy and other overtalkers from their talking? Not just the sheer amount of talking but also what people choose to talk about, how they talk about it, and what those choices may say about them? I want to know what we can learn about someone just by analyzing their talk.

For example, why did Podiatry Guy feel the need to tell me about some stranger having a disastrous foot surgery? In the absence of extensive shared history to draw on for conversation, people who don't know each other well sometimes emphasize negative information in conversation, likely because stories and warnings of disaster capture their audience's attention. This observation lends some insight into why some folks are constantly posting negative stuff on social media. They may just be the type who always thinks that the world is going to hell, but also they may be getting rewards for their doomsaying in the form of attention, likes, and clicks on their social media accounts. My hunch about Podiatry Guy is that he had a high need to talk, knew me only slightly, and he latched on to this attention-grabbing story to allow him to hold the conversational floor.

In this chapter, I'm going to show that we can do more than speculate about people like Podiatry Guy based on their talking patterns. I'll present the science of how speech and writing can not only reveal our tribe but also identify characteristics of individual talkers. I don't mean the completely obvious point that you can tell that someone loves old movies because they talk about old movies and how much they love them. Instead, we'll explore how a talker's particular word patterns can tell us other details about them. These details might include information that they don't intend to reveal, such as aspects of their personality and mental state, and whether they're lying.

Analysis of talking to uncover people's inner secrets might sound like the stuff of fiction, and indeed, language analyses have been plot points in plenty of movies and TV shows. As a psycholinguist who used to work at MIT, I have to give a special shout-out to one of these shows, the TV series *Dexter*. In one episode, the serial killer Dexter writes a fake manifesto to mislead the police, and MIT psycholinguists are called in to analyze the language patterns in the document, in hopes of identifying the author. Excitement ensues, but Dexter escapes detection yet again.

The *Dexter* plot roughly captures where we're going here, minus serial killers. We'll see real-world examples of psychologists, forensic linguists, and computer scientists working to glean information about people from their speech and writing patterns. In the United States and many other countries, these analyses are often based on publicly available writing, including transcripts of interviews and posts on Facebook, Reddit, and other social media. This kind of work once required poring over recordings and texts by hand, but now we have modern computers and analysis software, allowing for more sophisticated guesses about the people who produced this talk.

Arrangements of words and other talking patterns have the power to reveal something about the talker because of the way that talking works. As we've seen, our mental attention to the messages we want to convey is essential for retrieving the words we need from our long-term memory. More important concepts get more attention and tend to arrive from our long-term memory faster. Those early words in turn shape our sentence structure, which in turn guides our intonation if we're speaking, and our punctuation if we're writing. All of these external aspects of our talk—words, word order, grammar, intonation, punctuation—can then become data that allow researchers to work backward, estimating the internal state of the talker and the ideas they're focused on.

You can think of the telltale signs in our talking as similar to footprints in the snow—evidence that, in the right hands, can be really informative about the one who made the footprints. A nonexpert might see those tracks and be able to recognize that someone has passed by, but not much more than that. A skilled tracker might look at the same footprints and be able to estimate the walker's height, age, and other details that would escape the rest of us. Similarly, a skilled language researcher can notice patterns in a text and make much more educated guesses about the writer than nonexperts can. But when the tracker or the language researcher teams up with a companion who has complementary skills, there is an exponential increase in abilities to interpret clues, whether they're in snow or in talk. In the tracker's case, having a dog along provides a wealth of scent analysis, enough for the dog to follow and identify the walker. The researcher's companion is a computer program. Instead of scent, the program processes vast amounts of data, allowing it to detect talking patterns not evident to the naked ear or eye. Those subtle patterns turn out to be game changers in using talk to identify characteristics of the talker.

We'll ease into this topic with some older research that used the more limited computer programs of ten to fifteen years ago to examine samples of speech and writing that are fairly small by modern standards. The goal of this research was to investigate whether it's possible to detect evidence of cognitive impairment in someone's talk well before they are diagnosed with dementia. These analyses require looking for changing patterns in a body of recorded speech or writing gathered over many years. We'll see case studies of two people for whom extensive records of talk are available.

Talking and Mental Decline

Iris Murdoch was an illustrious British philosopher and writer who wrote twenty-six novels, many of them widely praised. Queen Elizabeth II made Murdoch a dame (the female equivalent to knighthood) in 1987, honoring her for outstanding services to British literature. Murdoch was diagnosed with Alzheimer's disease ten years later. Because the decline associated with Alzheimer's disease can precede a formal diagnosis by months or years, researchers analyzed her novels to determine whether a decline could be detected in her writing. They took samples of three of Murdoch's novels—her first one; her most famous work, written in the middle of her career; and her last novel, written a year or more before her diagnosis of Alzheimer's disease.

The second subject is Ronald Reagan, who was the fortieth president of the United States. Reagan was diagnosed with Alzheimer's disease several years after he left office. Again researchers were looking for evidence of decline before diagnosis, and they focused on his spontaneous speech, taken from his answers to reporters' questions during his forty-six press conferences while he was president.

For both Murdoch and Reagan, comparisons of talking patterns over time did reveal early evidence of cognitive decline, years before neurologists had made a dementia diagnosis. A major "tell" in both Murdoch's and Reagan's talking patterns was their changing use of words over time. Murdoch and Reagan both had fewer precise words in their later writing or speaking compared to earlier in their lives. Words that they could no longer rapidly retrieve from memory were replaced by common, vague substitutes that they did manage to find, words like *something*, *they*, and *this*. Both Murdoch and Reagan took good-enough talking to the extreme, where these vague words weren't really good enough anymore. Reagan's spontaneous speech in his later press conferences also included many more filled pauses such as *uh*, *so*, and *well*, compared to his earlier talking patterns. This increase in filled pauses also reflects more difficulty in getting words out of memory, because these little filler words are used to buy extra time for the talker until they can retrieve and plan other words.

The fact that evidence of mental decline can be found in talk patterns years before a dementia diagnosis is both alarming—Reagan was president during this decline—and also not that surprising. Talking is hard work requiring real focus and attention, and we've seen before that cognitive impairments have talking consequences. Indeed, when records of talking patterns over time are available, they could be a real asset to neurologists' cognitive tests for dementia diagnoses.

Talking and Mental Health

There is a serious mental health crisis in the United States and many other countries. About one in five adults in the United States has some form of mental illness such as anxiety, depression, schizophrenia, or bipolar disorder, and one in twenty has a condition so serious that it greatly interferes with their health and well-being. These illnesses place enormous burdens on the patient, their family, and the broader community.

Managing the health of mentally ill patients can be extremely challenging. Patients may refuse to go to a clinic or may not tell the truth to doctors or therapists. In the hopes of improving diagnosis and treatment, researchers are investigating whether analysis of a patient's writing, including publicly available posts to X/Twitter, Facebook, or Reddit, could contribute to developing a profile of a potential patient. If so, that extra information might be of real benefit to the patient and therapist, and to clinical practice more generally. However, someone wouldn't have to be clinically paranoid to worry that language analyses like this might be invading their privacy. We'll see both sides of these issues.

It's helpful to start with a hypothetical situation to get at your own intuitions. Imagine that researchers have permission to examine some individuals' talk, either their writing or transcripts of their speech, and the researchers use a language-analyzing computer program to assign the talk into two groups—talk that likely was produced by those with a diagnosed mental illness versus talk probably produced by people without mental illness. These modern computer programs are more powerful than the ones researchers used to analyze Murdoch's and Reagan's language, and they can potentially identify even subtler patterns that might elude humans who look at the language.

Mental illnesses, including depression, psychoses, and personality disorders, are not all the same, but for the purposes of this thought experiment, let's consider words or word patterns that might tend to identify any form of mental illness. Try writing down your guesses about which word patterns are informative about mental illness so that you can check your early intuitions against some of the major patterns that computer algorithms have discovered.

Now that you've got your list, I can tell you that your hunches are probably going to be completely incorrect. But you're in good company, because basically everyone doing this research started out wrong. This brings us back to Jamie Pennebaker from the University of Texas, who we last encountered in chapter 4. Back then, we saw his work on how expressive-writing exercises could help people process difficult situations in their lives. After studying the benefits of expressive writing for some time, Jamie realized that he could turn that research on its head to ask a different question: If writing can be a route to improving well-being, couldn't researchers analyze what people have written and get insight into their mental state?

That question led to Jamie effectively founding another field, modern computer-based language analysis. When he and his colleagues were just starting out, they had all sorts of ideas about what patterns might reveal well-being, mental illness, and so on. He told me that every one of his initial predictions turned out to be wrong. It was a shocker at the time, but Pennebaker is now quite gleeful about how naive he was. He loves the fact that his results show that simple, obvious ideas about talking don't actually hold water. The important language patterns weren't where researchers initially thought they'd be, and it took computer programs to reveal what was really going on in our talk.

Here's an example. You might have guessed that mentally ill people talk about very different topics than healthy ones, but choice of words doesn't distinguish groups of people very well. Words that carry a lot of meaning, like nouns and verbs, have what's called a *bursty* character in people's talk, meaning that the words burst onto the conversational scene for a few days or weeks in response to an event worth discussing. Maybe everyone in your hometown starts talking about some scandal in city hall that's come to light, and everybody is suddenly saying words and phrases like *payoffs*, *embezzled*, and *Mylar envelopes full of cash in the janitor's closet*. These expressions get used intensively for a while, but they fade away when folks' collective attention shifts to something else. Language analysis of these words in the townspeople's talk is great for telling us about trending topics in the town, but it can't tell us much about individual talkers, because many people are using these same words while the topic is trending.

If we're going to use word patterns to reveal something meaningful and relatively stable about individual people, we're going to have to look beyond the bursty words of the topics of the moment. Instead, computer programs look for common words that tend to occur across many topics. Some of these might be meaningful words like *sleep* or *stressed*, but there is another group of words that turns out to be much more informative: pronouns.

In English, pronouns, like *I*, *you*, and *it*, are sprinkled throughout our speech and writing. They are wonderful for efficient talking, replacing long phrases like *the head of the sales department* or *Ms. Washington* with something very short, *she*. Pronouns show up across almost every kind of talking and topic. These tiny handy words take center stage in studies of talking and mental health. Compared to people with no mental illness, those who have depression or anxiety use noticeably more pronouns referring to themselves—*I*, *me*, and *my*. This pattern is so robust that the use of these self-pronouns is routinely found to be the single best language predictor of whether someone has depression or not.

Did you list pronouns as a form of talking that could predict mental illness? Me neither. And neither did Jamie Pennebaker and his colleagues, who discovered this pattern. The self-pronoun and depression connection is a shocker for me because self-pronouns are already omnipresent in everyone's conversation. As I've mentioned before, *I* is the single most common word that young adults say in conversation. It's astonishing that computer algorithms can detect increased use of self-pronouns in depression amidst the backdrop of everyone already talking about themselves.

The extra-high use of self-pronouns in depression reflects an even greater attention to oneself in depression and anxiety. I don't mean this in an egotistical sense. Instead, depression can lead to turning inward, and this greater inward attention is reflected by tendencies for the depressed person to mention themselves and what they are worrying about and working through. Indeed, words related to thought—words like *think*, *believe*, *doubt*, and so on—also tend to be more common in the writing of people with depression than other people's writing. Putting self-pronouns and thinking words together, we can see that depressed patients are frequently trying to sort out their own mental state and situation, and these patterns are reflected in their writing—for those who know how to look.

Negative words like *sad* and *angry* sometimes appear more in the writing of depressed people than people with no mental health issues. You might have flagged these types of words as an obvious difference from healthy people's talk, but they're actually a very minor player. The pattern of increased negative-emotion words in depression isn't stable—it turns up in some studies of depressed patients' talk but not others. The explanation is likely in the nature of depression. We commonly use the word *depressed* to mean sadness, but the mental illness called clinical depression is associated more with a lack of normal emotion—an emotional emptiness more than a sadness. From this perspective, emotion words wouldn't necessarily pervade the talk of depressed people.

Can we put this new knowledge about word patterns in depression and use it ourselves to make inferences about someone's mental health? I really don't recommend it, because nonexperts don't have the training or computer tools to consider a wide swath of language. To make this point more concrete, consider this excerpt from an interview with Charles Bukowski, novelist, poet, columnist, and prolific writer of edgy portrayals of people scraping by in rough times. The interviewer, the actor Sean Penn, asks him about loneliness, and Bukowski replies, "I've never been lonely. I've been in a room—I've felt suicidal. I've been depressed. I've felt awful—awful beyond all—but I never felt that one other person could enter that room and cure what was bothering me . . . or that any number of people could enter that room. In other words, loneliness is something I've never been bothered with because I've always had this terrible itch for solitude."

Bukowski's reply seems like a winning ticket in a depression talk lottery. His answer is choked with self-pronouns, nine in total (*I* and *me*). He's got negative-emotion words going—there's *awful*, *suicidal*, and even the word *depressed.* And he's using thinking words, including *felt* and *bothered*. Bukowski himself explicitly says he had been depressed, and we shouldn't doubt him. But if the word *depressed* were missing from this statement, should we infer current depression from the patterns of his words?

No, I don't think so. It's important to consider the context here. Bukowski is being interviewed, being asked what he thinks about various topics. Anyone in that circumstance is going to have many self-pronouns and thought-related words. Those language patterns can simply be evidence of Bukowski being a cooperative interview subject, nothing more. And his negative-emotion words naturally follow from answering a question about loneliness—they're bursty words popping up in response to the introduction of a negative topic. Elsewhere in the interview, asked about more cheerful times, Bukowski expresses other emotions: "The sunlight was brilliant, and the sounds were great. I felt powerful, like a recharged battery."

This example shows the need to look for language patterns over a much larger expanse of talk to avoid patterns that emerge from a particular situation or topic. For this and other reasons, these analyses are promising but not yet used regularly in diagnosis of mental illness. Clinicians don't routinely have large samples of language handy to analyze, and they don't necessarily have the language analysis tools or experience that are available to language researchers.

I don't think that language analyses in mental health diagnosis and treatment will be out of reach forever. When they do become more widespread, we will have to consider the ethical consequences of mining patients' social media posts or other language for information about them. These analyses could yield significant loss of privacy, because they may suggest diagnoses or other personal information that the talker doesn't intend to reveal or even realize themselves. Research on the safety of electronic medical records shows that breaches of these records, while still rare, are becoming increasingly common. A breach of this information could affect someone's employment, insurance coverage, and more.

Many folks appear unconcerned about this potential loss of privacy, routinely accepting the terms-of-use agreements for social media companies, not realizing that accepting these terms may be allowing the companies and others to use their posts for research, diagnosis, or marketing, potentially for decades in the future. We'll see an example of research using social media next.

Talking Can Reveal Your Personality

Most personality researchers think of personality as a collection of stable traits or tendencies for how we engage with the world. They believe that they can often capture the core aspects of our personality by identifying where we land on five continuous trait scales—the Big Five, they're called. The most familiar of these five scales may be Introversion-Extraversion, where we humans run the gamut from extremely introverted to extremely extraverted. You probably have a good sense of where you and people you're close to fall on a scale of introversion/extraversion. I think I'm slightly on the introverted side, but not by much. My nonexpert guess is that Podiatry Guy is pretty high in extraversion.

The other four personality scales are Openness to Experience—the degree to which we seek routine versus want to try new things; Conscientiousness, extending from very impulsive at the low end of the scale to very thoughtful and planful at the high end; Agreeableness, reflecting the degree to which someone gets along with others; and Neuroticism—very moody and anxious at one end of the scale to very stable with few worries at the other end.

Researchers typically measure these five personality components with a long questionnaire asking the respondent how they'd behave in many different situations. This approach does a reasonable job placing people on the five scales—slightly extraverted, highly conscientious, and so on. Because personality differences affect how we view the world and presumably how we talk about it, it's natural to ask if there's another way to uncover someone's personality, via patterns in their talk.

In one of many studies addressing this question, researchers used data from years of Facebook status updates for over seventy-two thousand Facebook users. All of these Facebook users had taken a personality test on Facebook and given permission to have their answers used for research. The researchers didn't know the identities of the Facebook users, but they had all the users' personality test results, paired with their Facebook posts. The researchers developed and trained computer algorithms to predict personality scores from the language patterns in the users' Facebook status updates.

First the algorithm was trained to find connections between scores on the personality test and language patterns in the status updates for sixty-seven thousand of the users. Next the algorithm was given the status updates from five thousand other Facebook users and was asked to guess how these people scored on the Big Five personality scales, based on what the program had learned about talking and personality from the first set of Facebook users.

The talking patterns in Facebook status updates don't perfectly reflect the results of the personality tests, but they're very good. Considering that Facebook status updates are often just short posts about what people are doing in their day, and that the posts vary widely depending on events in the world, the success of these analyses is remarkable.

Prediction of someone's personality from Facebook posts sounds like a magic trick, and we naturally want to know what's going on behind the curtain. As with other analyses we've seen, there's no one pattern of talk that classifies personalities; the algorithm is making guesses based on many patterns—the words people use, word combinations, sentence length, even their punctuation. This "It's a combination of lots of stuff" answer isn't very satisfying to our human desire for clear examples of how the magic is done, so let's compare two different personality types, giving examples of some talking patterns that point toward one personality type or another.

People with the personality trait called Agreeableness are ones who tend to value social harmony and getting along with others. Those who score high on Conscientiousness tend to be responsible and organized, and they generally follow societal rules. Someone can be both highly conscientious and highly agreeable, but for the purposes of this example, let's look at the talking patterns of two imaginary people: Agnes, who scores high on the Agreeableness scale and average on Conscientiousness, and Connie, high on Conscientiousness and average on Agreeableness.

People like Agreeable Agnes and Conscientious Connie often look on the bright side in their Facebook status updates, using words like *happy*, *amazing*, and *wonderful*. Being very agreeable and being very conscientious are both positive traits, and so it's not surprising that Agnes and Connie tend to post about things being super fantastic. But they also have a striking difference: Agnes and other highly agreeable folks also tend to pepper their status updates with negative descriptions of people they disapprove of, using words like *stupid*, *pathetic*, *arrogant*, and *rude*. Conscientious Connie and others like her don't do this and in general don't have many negative words in their status updates.

The algorithm's job is to make personality predictions from talking patterns like these, not to speculate why personality and talking line up the way they do. I can make some guesses here about why these patterns of positive and negative words arise, but we can't verify or disprove them. One possibility is that people like Agnes and Connie all find rude people quite annoying, but Connie and other highly Conscientious rule followers don't want to violate social norms and call people out on Facebook. This tendency may help keep negative comments about other people out of their status updates, while Agreeable Agnes, who is only average in the conscientious trait, doesn't have such a strong rule-following inclination. A related possibility is that the Agreeable Agneses of the world, who so strongly value getting along with others, become absolutely furious at rude and disruptive people, and this frustration spills out in their Facebook posts. Meanwhile, Conscientious Connie, who is only average in the Agreeableness trait, is not so fixated on everyone getting along. She doesn't feel the need to dwell on these rude people when she's posting on Facebook.

We can't fully know what's driving these similarities and differences in talking patterns. But what we do know for sure is that for whatever complex cluster of reasons, our personality is reflected in our talking patterns clearly enough for strangers with a fancy computer program to know something about us, even if they've never met us and we're not talking to them. In the research I just described, the researchers didn't know the identity of any of the seventy-two thousand Facebook users whose data they were analyzing, but that anonymity isn't necessarily guaranteed in other situations.

The anonymity issue leads us to another question: What's the point of identifying someone's personality from their talk? It's not just an academic exercise; many companies and other groups would like to know this information. Often it's not that personality itself is so interesting, it's what personality signals about the real holy grail: predicting someone's behavior. Imagine what various groups could do if they had personality information about Agreeable Agnes, Conscientious Connie, and other Facebook users. The personality profiles could help companies predict who will be a good employee, who would be receptive to some kind of advertising, who might buy a car, who would be influenced by certain kinds of ads, and so on. They might send targeted ads or coupons on the basis of personality profiles, even if they don't know the identities of targeted users. Law enforcement could use personality information to guess who might engage in risky or illegal behavior. Political campaigns would use personality to guess who might vote a certain way and who might be persuadable by certain kinds of campaign messages. Information is power, and there's money and power to be gained from information that can be gleaned from our talk. I don't know whether personality-based analyses are part of the mix that companies are using when they target us with messages, but the possibility is real.

Of course, groups who are using language analyses might skip the personality assessment and go directly to trying to predict someone's behavior from the language they analyze. That's where we're heading next.

Liar, Liar, Talk on Fire

People sometimes lie. Maybe you've even tried it yourself. Of course you have, because everyone does at least some minor lying. Even the rule-following Conscientious Connies of the world toss out a few fibs. Maybe you've claimed that another engagement prevents you from going to your cousin's barbecue, or you've called in sick on a workday packed with boring meetings, or you've told your sister that you're not upset at her, not one bit.

Part of the reason that we produce these small-scale lies is that we can get away with them. That's another way of saying that the people we're lying to often aren't sure whether we're lying or not. But language analyses might not be so easily fooled.

Left to our own intuitions, we are terrible at detecting lies. Several careful studies of lie detection have shown that we're often no better than random guessing in distinguishing truth and lies in all kinds of situations. We can't tell the difference between genuine and fake opinions, real versus fake news, honest versus untruthful reports in crime investigations, and fake versus real reviews of merchandise posted online. Imagine for a moment that you're on trial for a crime, that witnesses are testifying, and that twelve earnest people in the jury box are doing their job, listening carefully. Unfortunately, those jurors may be incapable of figuring out which witnesses are telling the truth.

We're in big trouble if detectives, prosecutors, judges, and jurors can't tell a truthful witness report from lies, but the problems don't stop there. Fake news and other lies have led many people to try crackpot remedies for diseases or reject scientifically supported medical care, sometimes with fatal consequences. False advertising and scams routinely defraud people of their life savings. All over the world, government disinformation is used to keep citizens under control in authoritarian regimes and to influence the citizens of foreign countries, with the United States being a prime target of disinformation campaigns from overseas. It's no wonder that we can develop feelings of cynicism and helplessness around determining what's true, eroding our faith in science, medicine, government, media, and each other, even in situations in which objective evidence is available.

Several education programs have been launched that aim at training high school or college students to better identify false information. Some of these programs have failed spectacularly: In the process of exposing the students to false statements in the service of training them to distinguish truth and lies, the students don't just fail to improve; they get worse. The exposure to the fake information during training makes the fake stuff seem more familiar and therefore more true. At least with the programs available so far, we're not going to train ourselves out of this problem anytime soon. That's why some people look to technology for help in sorting truth from lies.

The first device to offer some improvement in lie detection was the polygraph machine. You've likely seen these in old movies and detective shows. An interviewer asks someone a series of questions, and physiological measures—the interviewee's heart rate, breathing, and so on—help the trained polygraph examiner to judge whether the person is lying or telling the truth.

Alas, many studies have concluded that polygraph-assisted lie detection is not highly accurate and is subject to biases by the human polygraph operator. That's why polygraphs are associated with old movies—they're less commonly used today. Polygraphs also have the disadvantage that the person being examined must be physically present, wearing all the measurement devices. That's unfortunate, because there are endless situations where we want to know the truth of something said on the phone, on social media, in a trial, in a public debate, or in a podcast or TV interview.

Computer analysis of transcripts of speech or writing might fill some of the need for better lie detection. These analyses don't require the talker to be present, and they don't require any special equipment to be worn by the talker.

There are two general ways in which a person's talk might reveal their truthfulness. The first idea is that truthful thoughts might slip out while someone is telling lies. This theory arose over a century ago, proposed by Sigmund Freud, the father of psychoanalysis. Freud believed that errors during talking could reveal our unconscious thoughts, as when someone accidentally says the truth, "I had to kill them," when they had intended to lie and say, "I had to save them." If Freud is right about unconscious thoughts leaking into our talk, we might look at errors like these for signs of lying.

Well, it's not that simple. Applying what we know about how we plan our talk, this "kill them" error likely reflects the word *kill* being pulled out of long-term memory instead of *save*. Yes, this mistake in memory retrieval might indicate the talker's failed intent to suppress the fact that they have killed someone. But other possibilities are also likely. *Kill* might have come up from memory because the person contemplated saying the truthful statement "I had to save them from being killed." Or the error might arise for some other reason; talkers make word substitution errors like these all the time, without any hidden thoughts. If I accidentally say, "Turn right," instead of "Turn left," or "I'll chop the onions," when I mean that I'm chopping the garlic, I'm not lying or revealing any unconscious obsessions with garlic or left turns. Errors like these happen regularly because the process of retrieving just the right words from memory is difficult and imperfect. Mistakes happen, and we can't use them as clear evidence of lying.

If we can't rely on errors to show us the truth seeping out of our talk, we need to explore a second method, relying on the fact that talking is hard work. Remember that in chapter 3 we saw that it can be quite challenging to talk while doing something else, such as driving or tracking a moving dot on a computer screen. Talking while lying is also doing several things at once: Someone who's telling a story truthfully is using a memory of real events to drive their words and other aspects of planning what to say, but a liar is simultaneously planning their talking while also trying to make up something they didn't experience and hide what they want to keep hidden.

The liar's extra difficulty can be reflected in their talk. The liar's story, which drives the words they choose, is likely going to be less detailed than when someone speaks truthfully, because the liar doesn't have the real memory to draw on. The liar may also be trying to suppress reference to some real events but mix other events into their story, all while generally trying to sound like they're telling the truth. Talking while lying makes for a cocktail of difficulty, and the consequence may be that a liar's talk is not like the truth teller's.

Researchers have tested these ideas by randomly assigning people to either tell the truth or lie about something—some event that happened, their opinion on some controversial issue, and so on. The researchers then analyzed patterns in speech or writing to search for features that distinguish the people told to lie from the ones who were told to tell the truth. In some cases, other folks were asked to read the statements and try to identify which ones were true and which were lies.

These studies have turned up several interesting results. Once again, human judges weren't any better than random guessing at detecting true versus false statements. Computer analyses of the language of the truthful and lying talkers did better. The language analyses assigned two-thirds of the written and spoken statements correctly into "truth" and "lie" categories. That's still far from perfect, but it's much better than we can do on our own.

As usual, the computer analyses are making their guesses based on a large pile of cues that they uncover, but we can look at some of the major patterns. Overall, truthful texts had more complex language than the lies. This difference played out in several ways, including the true statements having more abstract thought-related words, such as the verbs *think* and *believe.* The true statements also had more words indicating nuance in beliefs, with words like *except* and *without.* False statements tended to have very common verbs like *go,* and little discussion of the nuances of any opinion. A liar's fake message is sketchy compared to someone's real report, and the fake message can't generate rich descriptive words, leaving the liar's story limited to generalities.

Researchers are trying to take these controlled experiments, in which they know who is lying and who is not, out into the real world, to address issues such as detecting fake news. Fake news is typically written by professional fake-news writers rather than everyday people instructed to lie, as in the study we just saw. Fake news might therefore be harder to detect than lies generated on command in an experiment. The difficulty in spotting fake news is a serious problem. Some folks are persuaded by even provably false statements, and probably all of us are tempted to believe some fake news, given our overall low skill in detecting lies.

Let's try out your intuitions! Here are some political news headlines, all taken from studies of fake-news detection conducted during Donald Trump's first term as president of the United States. The researchers used headlines about Trump because presidents are naturally prominently in the news, and Trump routinely acted in ways that were atypical compared to past presidents. Together, that made a good environment for a study on real versus fake-news reports about Trump's behavior and administration.

The headlines below might all be from fake news, might all be from real news stories, or some mix. What are your guesses, and how certain are you about each of them?

1. Facebook Removes Trump Ads with Symbols Once Used by Nazis
2. President Trump Fires All 14 Muslim Federal Judges
3. Trump Science Advisor Denies Apollo Moon Landings Ever Happened
4. White House Staffers Defect, Releasing Private Trump Recording

If you had difficulty deciding, you're not alone. The research participants who viewed these and similar headlines rated the ones from real articles on average as being somewhat more likely to be true than the fake ones, but the raters were uncertain about the true/false status of many of them. If you identified one of these headlines as real and three as fake, you're on the right track. The real one is the first one, about Facebook ads, and the rest are fake.

Of course, an isolated headline isn't much to go on, and we can also look at fake-news detection using entire news articles. Computer algorithms trained to distinguish real and fake news have found that the fake reports tend to be more surprising than real news. Most real news has some sameness to it—a new law was passed, the on-ramp to Highway 12 is closed for road repairs, travelers are experiencing holiday flight delays, same old stuff. Real journalists file reports on important events even if they are not particularly surprising, but fake news is designed to grab our attention. On average, fake news is more surprising and unusual than the real thing. Whereas unprofessional liars produce lies that are vague and neutral, the professional lies of fake news go for the attention-grabbing jugular.

Fake-news creators also include emotional commentary that manipulates the reader's emotions. All of this interesting, emotionally engaging, and fake text can be churned out faster than real news can be reported, because fake news doesn't require time-consuming editing, fact-checking, or other activities of real journalism. It can even be generated by artificial intelligence systems that have been trained on the language patterns in fake-news stories. Fake news that includes photos also tends to have better pictures than real news, because while real news photographers are limited by the real people or scene related to the story, fake-news generators can scour stock photo sites or manipulate photos to get something eye-popping.

The high levels of interest in and emotional engagement with fake news are undoubtedly part of the problem that fake news poses. A study of fake news on X (formerly Twitter) showed that it's retweeted and reposted far more often than real news, spreading six times as fast by some measurements. And we can't blame automated bots for this problem; the rapid spread of fake news appears to be due to humans retweeting it, not computerized bots. And remember what we learned about hate speech—the more often someone produces an opinion, the more strongly they believe it. We can expect the same about repeating and reposting fake news.

I'd like to be able to tell you that high-powered computer analyses are powerful allies in the fight against fake news, but the truth is much more sobering. There is intensive ongoing research on computer models for fake-news detection, and the field is rapidly evolving, but as I'm writing this, the many approaches being tested are not really doing that well. Good detection of fake news requires not only knowledge of language patterns but also awareness of events going on in the world, which change every day. And fake news appears in many different venues with different characteristics—"news" articles, Facebook posts, blog posts, discussions on podcasts, fake videos, and more. Different approaches may be needed for different types of fake news. And alas, there is so much of it, and it spreads so quickly.

Don't Try This at Home

Having read this far, you may be wondering whether you could use some of the information revealed by computer analyses of language to improve your own interactions with people or media. The computers are crunching enormous amounts of data and weighing thousands of possible cues. That's more than we humans could ever detect, especially in the middle of a conversation. But could you make meaningful gains in your analysis of others by attending to just a few of the major talking patterns?

Maybe. Remember the Charles Bukowski interview we discussed earlier, and how difficult it is to make inferences from small snippets of talk? I think your best bet is to use the information reported here to turn a skeptical eye to possible fake news—compared to real news, the fake stuff is more surprising, interesting, and emotionally engaging, with more pretty pictures and more viral spread. Fake news is hard to detect, but folks who are more analytic—pausing to think about what's plausible versus what's illogical—are better at spotting it than those who just go with their gut instinct. Someone who combined this analytical approach with the knowledge that fake news tends to be surprising and emotional probably could acquire some additional edge in fake-news detection.

It's certainly true that observant humans have long been able to detect some language patterns without the aid of computers. Jane Austen, an absolute expert in how talk patterns could reveal a character, built those insights into the dialogue of her novels to great effect. Here is young Harriet Smith in the novel *Emma*, telling Emma about her recent encounter with a young man she fancies, over Emma's objections:

> *I found he was coming up towards me too—slowly you know, and as if he did not quite know what to do; and so he came and spoke, and I answered—and I stood for a minute, feeling dreadfully, you know, one can't tell how; and then I took courage, and said it did not rain, and I must go; and so off I set . . .*

Harriet's tale oozes awkwardness, uncertainty, and anxiety. She tells us directly that she doesn't know what she's feeling, but we can also see her unease and bewilderment from the frequent addition of *you know*. That filled pause is rather like the modern *like*: it buys Harriet extra time to figure out what she wants to say. And her description lists one little event after another, suggesting that Harriet is just unloading her memory in minute detail, unable to separate the important from the trivial. If you've listened to teenagers narrate recent events in their own lives, this stringing along of little micro-moments will seem very familiar.

Jane Austen used a lifetime of keen observation to write passages like this. It's possible you could get similarly good at detecting language patterns, but it would be a major undertaking. And no one observer is going to surpass the computer-based number crunching we've reviewed here.

Rather than taking what you've learned in this chapter and attempting more assessments of people, maybe the lesson is to go in the opposite direction. The difficulty of identifying much about a person from a small sample of talk makes me reconsider the kinds of snap judgments I might make about talkers. My guesses about Podiatry Guy—overtalker, likely extraverted, eager to tell me scary stories about foot surgery to capture my attention—stem from a truly small sample of data and might be way off base. And remember what the language analyst Jamie Pennebaker said—when he and others first started analyzing texts to identify depression, personality, and so on, EVERY SINGLE ONE of their initial hypotheses was wrong. Best to suspend judgment if all we have is skimpy data and our own possibly faulty intuitions.

What we do get from this discussion of computer-aided analyses of talk is a sense of promise and pitfalls for the future. The computer-aided analyses are doing quite well and are going to continue to get better. They're going to be applied more broadly. They might have trouble with fake news for some time, but maybe they'll be better at predicting which people who are spouting hate speech are most likely to act on their words. If talk patterns could help law enforcement predict probable future hate crimes, those predictions could possibly be the basis for important crime prevention attempts, including using the predictions to support legal injunctions against the hate talker, barring them from contact with the group they're targeting, or using red flag laws that revoke the hate talker's right to own guns.

And better diagnosis of mental health may also be on the horizon. Given the need and the difficulty of mental health treatment, help from talk analysis could be a real win. And for middle-aged or elderly adults, collecting talking samples could be incorporated into doctors' visits over the years, with the goal of amassing talking data to use for early detection of cognitive decline in later years. As more memory drugs come on the market, earlier detection aided by talk samples could make a huge difference in minimizing cognitive decline.

Despite these potential benefits coming down the road, you do want to think about your views on having your own talk analyzed. How do you feel about the lack of privacy you're inviting when you sign up for anything that might be tracking and storing your talk? A majority of Americans believe that social media companies are not sufficiently protecting their privacy. Probably most of them are thinking about data breaches in which private information is collected by an unauthorized group, such as hackers collecting bank account information or passwords. That's definitely a concern. But we should also recognize that our talk is being analyzed to reveal information about us, and our acceptance of user agreements has likely given companies free rein to do so.

Given the almost daily advances in artificial intelligence and the current rush of products to market, we really don't know what the upper limit will be on the success of talk analyses. And we also don't know the upper limit on where they'll be applied and how much they will reach into our lives. In the wrong hands, these algorithms could be used to attack free speech or discriminate against those who are legitimately questioning authority.

Collectively and individually, we should keep informed about both the promise of this technology and the safeguards that are needed to prevent abuse.

Afterword

There's a common piece of writing advice that's sometimes misattributed to the Russian novelist Nikolai Gogol. It's designed to push writers to leave a lasting impression on their readers, and it goes like this: The true, unwritten last line of every story is "And nothing was ever the same again." By encouraging young writers to think about whether their stories could merit this unwritten last line, writing instructors are prompting their students to create a meaningful transformation in their main character. By the end of the story, whether the character recognizes it or not, the reader should be struck by this change and should think about this character in a new light.

It is surely hubris on my part to wish that this book could make nothing ever the same again for my main character, talking. For nine chapters, I've been pursuing an argument that the nature of the mental processes converting ideas to talk have shaped our lives in unexpected ways. I've hit you with perhaps too many sentences along the lines of *We get this benefit because of the way that talking works* . . . because I wanted to be clear that these benefits come from our brain's efforts to plan our talk and manage the difficult task of talking. These talk-planning processes include focusing our attention to dig words out of long-term memory; assembling sentences just in time; starting sentences easy-first by beginning with whatever easily found words are available; cutting corners to make talking good enough by using common, vague, and ambiguous words; and cutting more corners by reducing our articulation when we can. If talk creation somehow worked differently in our brains, talking might have affected our lives very differently. And if we didn't talk, then all those benefits would vanish, and truly nothing about how we think, perceive, and act would ever be the same.

I'd like to believe that these arguments about the nature of talking and its effects could give talking its moment, where policymakers work to bring the benefits of talking into our schools, and where we consider how to include talk-focused activities to enrich the lives of children and elderly adults. I'd like to see the value of self-talk, already well recognized in sports psychology, become more broadly discussed in other areas where we know it influences our behavior. These influences include effects as different as emotion regulation, improved reflection and well-being, and increased learning. And I hope you now have insight into how we all bring biases to our own ways of talking, which lead to strife between us and others.

Though I have these hopes, we know from chapter 5 on talking and learning that it's incredibly hard to build lasting learning from comprehension alone, whether you're comprehending a book on talking or anything else. Because new knowledge fades away unless we do something with it, I'm offering some suggestions here for what you might do with your new knowledge of how talking works and the benefits that follow:

Reframe old ideas about talking

- Don't make a habit of finishing other people's sentences. Yes, sometimes you know what someone else is going to say. That doesn't make you unique; just about everyone can predict upcoming words sometimes. It simply stems from the fact that talking runs slower than comprehending, and sometimes when we're comprehending someone else, we use the extra time to predict their words. It's OK to pipe up with some next words occasionally, but no one likes a compulsive sentence-finisher.
- Think about how you might let go of prejudices about the way other people talk. Separate groups will naturally have different ways of talking. There are certain social pressures encouraging us to talk a certain way, wear certain clothes, and so on, but these are social conventions, not judgments about what's inherently better or worse. Creative types who push back against social norms are often celebrated. If we can admire those iconoclasts, we can also recognize that other societal variation should also be valued, including variation in the way we talk.

- In your youth, you irritated the adults in your midst by talking in newfangled ways. Now, when you encounter those younger than you, recognize that they are continuing down that same innovative talking path that you pursued in your own youth. You may feel uncool not knowing what they're saying, but their new talk is not inherently worse. You can't stop language change anyway, so don't make enemies of the young folks by complaining about it.

- Recognize that your own special pattern of talking really is unique to you, your personality, your interests. Think about what you allow media companies to record about your talking, such as your posts and comments on social media apps.

Use talking to focus, learn, and maintain cognitive skill

- Don't feel bad about talking to yourself, and don't make fun of anyone else who talks to themselves. It's valuable for keeping yourself on track, whether you're on the track running a race or anywhere else you need focus and follow-through.

- Indeed, when you need to learn something new, lean into talking. Tell the new information to someone else, write it down, say it to yourself. Use talking again later to test yourself and see if you're really laying down the long-term memories you need for learning. Resist the urge to go the easy route of just rereading the information, which leads to the feeling of learning more than actual learning. Exploit the desirable difficulty of talking to learn better.

- Don't spend money on brain-training games unless you find them fun. Even the ones related to talking—making words from a set of letters, word searching, and so on—don't currently have good evidence behind them that they lead to sharper thinking or memory. (The exception is crossword puzzles, which do have some evidence for benefits.) It's fine to enjoy these so-called brain games, but they're not a substitute for the stimulation of conversation and new learning in old age. Perhaps for someone who has little or no conversation options, these games could provide a little talking-like stimulation, but we don't know that for sure.

- A colleague who studies video games often gets asked which games people should play to keep mentally sharp. He's got a great answer: Play a game that you didn't play yesterday. Variety in activities, including games, is important mental stimulation. Old age, with its narrowing social networks and activities, can drain away variety. Seek out new activities, and then talk about them too.

- In Western cultures, where families are scattered and older adults often live far away from children and grandchildren, there is a serious shrinkage in conversation opportunities in retirement and with additional aging. Each family has different constraints and wishes, but paths that lead to more social interaction are beneficial for elderly adults, and often for everyone.

Engage kids in conversations

- Don't let other people dissuade you—babytalk is helpful for your baby and for adult-baby social interactions. It will naturally fade away when it's no longer beneficial.
- Parents, daycare providers, and anyone who interacts with little kids should understand that when infants are pre-talking at them or a toy, they're actively trying to engage the adults' attention. Adults can't respond to every *bababa* that the child produces, but frequent engagement is an important foundation for children's language, cognitive, and social development.
- Talking to children should include interactions that get the child talking and conversing, as the child's age allows. The back-and-forth of conversation is important for child language development, which in turn promotes children's executive functions—their attention, memory, focus, and readiness for school. Remember that they're not sponges; they need to act and engage and talk with others.
- Almost all of this important back-and-forth with young children disappears if a kid is focused on a phone or tablet. And it's equally sabotaged by having the adult on a phone or tablet. We love our media, but you should figure out arrangements in your family that promote conversation among everyone.

- A great route to kid-adult conversation is reading books with a child. Reading the words as they're written is a valuable introduction to book language, which is quite different from speech and is important for learning to read. Pausing to discuss what's going on in the book is also valuable conversation, including giving children turns to talk. Do each as the book allows.

- Learning to read takes significant practice outside of school, and there's evidence that kids are reading less outside of school than in previous years. Help your kids with this practice by having them read aloud every day during the first several years of elementary school. More practice yields not only more reading skill but also better school success and more enjoyment of reading in later years.

- When you or your kids have an assignment to do a presentation or write an essay, recognize that these exercises are not simply about engaging with whatever the topic is. Assignments like this are also honing writing and other talking stills that have real value in life. Convey the importance of these assignments to kids.

Talking, emotion, and mental health

- If you're upset about something, try pausing to consider exactly what emotions you're feeling. Naming emotions, particularly negative ones, is effective in bringing clarity and helping you not go overboard.

- If you have a complex decision to make, such as whether to move or take a new job, use the act of writing to help you think through the alternatives. Yes, write down the pros and cons, but also try to articulate how meaningful they are to you. There's no guarantee with any writing exercise, but there's plenty of evidence that these activities can be helpful in clarifying your thinking.
- Also consider writing down your thoughts about long-term issues in your life. Writing can help you gain insight about what is upsetting, get a better sense of your own emotions, help to put old frustrations or traumas into a broader perspective, and perhaps see a path forward.

Having written this book, I'm reminded of that Joan Didion line "I write entirely to find out what I'm thinking." Writing has shown me more of what I'm thinking, and these points about talking have become clearer to me. I've even made some steps toward more talking myself. No, I don't buttonhole people to regale them with tales of that foot surgery I had a few years ago. I'm not aiming to become an overtalker, but I am trying to be more willing to reach out to get together with friends, tell someone something interesting, and generally not wait quite so much for others to engage with me. As with all other aspects of talking, this part gets easier with practice. I'm looking forward to practicing more.

Also, having written a book about talking that's not about communication, I can't help but think about all the wonders of communicating that I've left unsaid. The benefits of conversation don't stop at the benefits for the talker. Conversing really is the best route for us to bond with each other and have lasting and meaningful social relationships. This book is not about communication, but communicating is a fantastic part of talking too. Perhaps there's someone you've been meaning to talk to but haven't connected with recently? Maybe you should figure out a way to get together with them and do some talking.

Acknowledgments

BACK IN 2019, WHEN STARTING THIS BOOK SEEMED LIKE A near-impossible goal amidst my day job of being a professor, I spotted an announcement for a summer book writing fellowship at the University of Wisconsin–Madison's Institute for Research in the Humanities. The program was designed for humanities scholars writing their second scholarly book. It provided a weekly writing group for authors to talk about their progress and some funds to pay consultants for feedback on a book proposal or early chapter drafts. I wasn't a humanities scholar, I wasn't working on my second book (this is my first), and I wasn't planning a scholarly work—my book was aimed at general audiences. Even though I didn't fit any of these criteria, something made me decide to apply anyway. Incredibly, the institute director, Steve Nadler, accepted me into the program. The weekly writing group helped me feel and think like a book writer. I acquired two invaluable writing buddies, Jennifer Ratner-Rosenhagen and Laura McClure, who provided encouragement, friendship, and writing accountability. My outside consultants, Julie Sedivy and Mark Winston, read my early writing attempts and gave invaluable advice that carried me through the whole process. All these experiences were transformative, and I am so grateful that Steve Nadler gave the book this early push.

I greatly benefited from talking with, and in some cases interviewing, colleagues near and far, all of whom improved this book. Thank you, Markus Brauer, Nick Buttrick, Dustin Chacón, Morteza Dehghani, Jeremy DeSilva, Maksim Hanukai, Judy Harackiewicz, Diana Hess, Nour Kteily, Paula Niedenthal, Jamie Pennebaker, Seth Pollak, Jenny Saffran, Dan Weiss, Mark Winston, and Klaus Zuberbühler. Though they're responsible for making many parts better, any errors are mine alone.

Many generous people read some or all of this book and offered encouragement and wise counsel. Thank you, Anya AitSahlia, Amit Almor, Vic Ferreira, Elaine Loring, Jessica Montag, Kate Nation, Karalyn Patterson, Ethan Seidenberg, Mark Seidenberg, Athena Skaleris, Bob Slevc, Chuck Snowdon, Gail Yaffe, and Michael Yaffe. They too made the book better.

My agent, Eric Lupfer of United Talent Agency, provided wonderful advice and encouragement, beginning with the new-to-me task of writing a book proposal all the way through the final stages of production. And huge thanks to my editor at Avery, Lucia Watson, who saw the potential of a book about talking that wasn't about communication. Lucia was endlessly supportive of my telling this story and helped me to tell it more clearly. Isabel McCarthy and the rest of the Avery team have been fantastic in designing the look of the book and shepherding it through the many stages of publication.

My graduate students and postdocs and the research we've done together have been a source of tremendous pride, in addition to being inspiration for parts of this book. I am lucky to have worked with Dan Acheson, Amit Almor, Mike Amato, Silvia Gennari, Arella Gussow, Todd Haskell, Elise Hopman, Yaling Hsiao, Cassandra Jacobs, Mark Koranda, Jelena Mirković, Jessica Montag, Neal Pearlmutter, Steve Schwering, Lynne Stallings, Karen Stevens Dagerman, Robert Thornton, and Justine Wells. Wonderful graduate student and postdoc collaborators from other labs include Morten Christiansen, Jon Willits, and Martin Zettersten. And I'm extremely grateful for all the undergraduate and graduate students in the classes I taught. Their insights and questions were critical for my own thinking and for understanding how to tell some of the stories here.

Several very different institutions have supported me over the years, for which I'm very grateful. Research requires funding, and I appreciate the support that students and I have received from the National Science Foundation, the National Institutes of Health, and the University of Wisconsin–Madison. And I also appreciate the Madison Public Library, where I've spent many hours writing, especially in the Sequoya library branch, named for the developer of the Cherokee writing system.

I was lucky enough to have two influential grandmothers in my life, and my cousins Roddy Coles and Ann Fillback Riley were very helpful in checking my memory for some of the grandmothers' remarkable qualities. My parents, Janet and Bill MacDonald, were my earliest writing role models and established the idea that writing well is a virtue. My mother, though she was an English teacher and not a psychologist, was shockingly observant about human and animal behavior. I believe that I became both a researcher and someone interested in language in large part because of her. I wish she and my father were around to read these thanks.

And saving the most important for last, as the writing guides recommend: My children, Claudia and Ethan Seidenberg, are an absolute joy. They're also a source of abundant interesting ideas on many topics, including talking. And very, very last, my husband, Mark Seidenberg, who makes me laugh and makes me think, often at the same time. I have been so lucky to have him in my life and to have gotten his ideas into my brain. This book could never have been written without him. Thank you, Mark, for everything.

NOTES

Chapter 1: We Do the Talking

3 **Doctor Dolittle books:** Some are available from Project Gutenberg, including the first one: Hugh Lofting, *The Story of Doctor Dolittle: Being the History of His Peculiar Life at Home and Astonishing Adventures in Foreign Parts* (Frederick A. Stokes Company, 1920), https://www.gutenberg.org/ebooks/501. Despite their adaptation into several films, the books haven't aged well. They contain racist depictions of non-Europeans, and there generally are very few interesting female characters. But the talking animals sure were fun.

4 **The comparison between animals' ability:** A note on terminology here. Biologically, humans are animals, primates, and apes, but I will use the term *primates* to refer to nonhuman members of the primate family, *apes* to refer to nonhuman apes, and *animals* to refer to nonhuman animals.

5 **A Diana monkey sits:** I first learned about Diana monkeys from Olivia Judson's wonderful science blog in *The New York Times*. Olivia Judson, "'Leopard Behind You!,'" *New York Times*, October 6, 2009, https://archive.nytimes.com/opinionator.blogs.nytimes.com/2009/10/06/leopard-behind-you/.

5 **They have calls:** Klaus Zuberbühler, "Referential Labelling in Diana Monkeys," *Animal Behaviour* 59, no. 5 (May 2000): 917–27, https://doi.org/10.1006/anbe.1999.1317. This video shows Diana monkeys, Campbell's monkeys, and other species giving *leopard* calls in the presence of a fake leopard: "Terrified Monkeys Scared of Leopard!," *Attenborough: The Life of Mammals*, posted February 8, 2013, by BBC Earth, YouTube, https://www.youtube.com/watch?v=o-bxPLFt1vI.

7 **In one experiment, Klaus:** Klaus Zuberbühler, "Interspecies Semantic Communication in Two Forest Primates," *Proceedings of the Royal Society B: Biological Sciences* 267, no. 1444 (April 7, 2000): 713–18, https://doi.org/10.1098/rspb.2000.1061.

7 **Yellow-casqued hornbills:** Hugo J. Rainey, Klaus Zuberbühler, and Peter J. B. Slater, "Hornbills Can Distinguish between Primate Alarm Calls," *Proceedings of the Royal Society B: Biological Sciences* 271, no. 1540 (April 7, 2004): 755–59, https://www.ncbi.nlm.nih.gov/pmc/articles/PMC1691652/.

8 **Both eagles and leopards need surprise:** Klaus Zuberbühler, Ronald Noë, and Robert M. Seyfarth, "Diana Monkey Long-Distance Calls: Messages for Conspecifics and Predators," *Animal Behaviour* 53, no. 3 (March 1997): 589–604, https://doi.org/10.1006/anbe.1996.0334.

9 **There's a long list of species:** This article sums up some of what happened in the animal language training field. Irene M. Pepperberg, "Animal Language Studies: What Happened?," *Psychonomic Bulletin & Review* 24, no. 1 (February 2017): 181–85, https://doi.org/10.3758/s13423-016-1101-y.

10 **A border collie named Chaser:** John W. Pilley and Hilary Hinzmann, *Chaser: Unlocking the Genius of the Dog Who Knows a Thousand Words*, reprint edition (Mariner Books, 2014). Here's a video of Chaser: "Chaser the Dog Shows Off Her Smarts to Neil deGrasse Tyson," posted August 20, 2018, by NOVA PBS Official, YouTube, https://www.youtube.com/watch?v=omaHv5sxiFI.

11 **In a video linked in the endnotes:** The nonsense symbols on Kanzi's keyboards are called "lexigrams." "Kanzi with Lexigram," posted November 8, 2007, by Iowa Primate Learning Sanctuary, YouTube, https://www.youtube.com/watch?v=wRM7vTrIIis.

12 **Kanzi did fairly well:** This video shows Kanzi following unusual commands: "Kanzi and Novel Sentences," posted January 9, 2009, by Iowa Primate Learning Sanctuary, YouTube, https://www.youtube.com/watch?v=2Dhc2zePJFE. See also: E. Sue Savage-Rumbaugh et al., "Language Comprehension in Ape and Child," *Monographs of the Society for Research in Child Development* 58, nos. 3–4 (1993): i–252, https://doi.org/10.2307/1166068.

12 **In Franz Kafka's "A Report for an Academy":** Franz Kafka, "A Report for an Academy," trans. Ian Johnston, Franz Kafka Online, accessed March 4, 2023, https://www.kafka-online.info/a-report-for-an-academy.html. Red Peter refers to himself as an ape, but he had a tail, and apes don't have tails. Kafka may have been thinking of Red Peter as a Barbary macaque, a large monkey (not ape) species that was called a Barbary ape in Kafka's time. This article by the leader of the Nim project also discusses this Kafka story: H. S. Terrace, "Apes Who 'Talk': Language or Projection of Language by Their Teachers?," in *Language in Primates: Perspectives and Implications*, ed. Judith De Luce and Hugh T. Wilder (Springer, 1983), 19–42, https://doi.org/10.1007/978-1-4612-5496-6_2.

13 **Language so dominates our lives:** Evelina Fedorenko, Steven T. Piantadosi, and Edward A. F. Gibson, "Language Is Primarily a Tool for Communication Rather Than Thought," *Nature* 630, no. 8017 (June 2024): 575–86, https://doi.org/10.1038/s41586-024-07522-w.

15 **When infants make these sounds:** Ed Donnellan et al., "Infants' Intentionally Communicative Vocalizations Elicit Responses from Caregivers and Are the Best Predictors of the Transition to Language: A Longitudinal Investigation of Infants' Vocalizations, Gestures and Word Production," *Developmental Science* 23, no. 1 (2020): e12843, https://doi.org/10.1111/desc.12843.

16 **Despite years of training:** Video of Viki saying some of her words after years of training: "Chimp Learns How to Speak the Word 'Cup' (from 'The Alphabet Conspiracy')," posted December 2, 2011, by ddeennnnyy, YouTube, https://www.youtube.com/watch?v=V7QM97fnypw.

16 **We now understand several reasons:** This article discusses both vocal tracts and brain networks: W. Tecumseh Fitch, "The Biology and Evolution of Speech: A Comparative Analysis," *Annual Review of Linguistics* 4, no. 1 (2018): 255–79, https://doi.org/10.1146/annurev-linguistics-011817-045748.

18 **The chimpanzee Nim, who was trained:** Nim has been the subject of many scholarly articles but also popular journalism and a heartbreaking documentary, *Project Nim*: James Marsh, dir. 2011. *Project Nim*. Documentary. Red Box Films, Passion Pictures, BBC Film.

19 **The researchers who trained Kanzi:** E. S. Savage-Rumbaugh et al.,

"Language Comprehension in Ape and Child," *Monographs of the Society for Research in Child Development* 58, nos. 3–4 (July 1993), i–252, https://doi.org/10.2307/1166068.

19 **One theory, developed by primate researchers:** Julia Fischer and Tabitha Price, "Meaning, Intention, and Inference in Primate Vocal Communication," in "An Overview of Nonhuman Primates' Communication and Social Abilities through Behavioral and Neuroscientific Approaches" (special issue), *Neuroscience & Biobehavioral Reviews* 82 (November 2017): 22–31, https://doi.org/10.1016/j.neubiorev.2016.10.014.

20 **The second theory:** M. S. Seidenberg and L. A. Petitto, "Communication, Symbolic Communication, and Language: Comment on Savage-Rumbaugh, McDonald, Sevcik, Hopkins, and Rupert (1986)," *Journal of Experimental Psychology: General* 116, no. 3 (1987): 279–87, https://doi.org/10.1037/0096-3445.116.3.279.

23 **The oldest known cave art:** Pallab Ghosh, "World's Oldest Cave Art Found Showing Humans and Pig," BBC.com, July 3, 2024, https://www.bbc.com/news/articles/c0vewjq4dxwo.

23 **There may eventually be DNA:** Fitch, "Biology and Evolution of Speech."

23 **Some have suggested that informing:** Szabolcs Számadó, "Pre-Hunt Communication Provides Context for the Evolution of Early Human Language," *Biological Theory* 5, no. 4 (December 2010): 366–82, https://doi.org/10.1162/BIOT_a_00064.

24 **These sites show fossilized footprints:** Two articles with evidence about group childcare include this Neanderthal site: Jérémy Duveau et al., "The Composition of a Neandertal Social Group Revealed by the Hominin Footprints at Le Rozel (Normandy, France)," *Proceedings of the National Academy of Sciences* 116, no. 39 (September 24, 2019): 19409–14, https://doi.org/10.1073/pnas.1901789116; and an older African site: Flavio Altamura et al., "Archaeology and Ichnology at Gombore II-2, Melka Kunture, Ethiopia: Everyday Life of a Mixed-Age Hominin Group 700,000 Years Ago," *Scientific Reports* 8, no. 1 (February 12, 2018): 2815, https://doi.org/10.1038/s41598-018-21158-7.

Chapter 2: Talk Baby to Me

27 **Dogs also don't talk:** This link shows the frequency of the singular and plural words for *baby, dog, infant* in books written in English (for adults or children) since 1800. Adding up the *dog* words and the *baby/infant* words, the dogs win out in many decades. The singular word *baby* begins to edge out *dog* about forty years ago, which might reflect the greater number of books for babies published in recent decades. https://books.google.com/ngrams/graph?content=baby%2Cdog%2Cbabies%2Cdogs&year_start=1800&year_end=2019&corpus=en-2019&smoothing=3.

29 **They do their reps:** Yuna Jhang and D. Kimbrough Oller, "Emergence of Functional Flexibility in Infant Vocalizations of the First 3 Months," *Frontiers in Psychology* 8 (March 24, 2017), https://doi.org/10.3389/fpsyg.2017.00300.

29 **These aren't fixed vocal routines:** D. Kimbrough Oller et al., "Functional Flexibility of Infant Vocalization and the Emergence of Language," *Proceedings of the National Academy of Sciences* 110, no. 16 (April 2, 2013): 6318–23, https://doi.org/10.1073/pnas.1300337110.

30 **The opening moves:** V. P. S. Ritwika et al., "Exploratory Dynamics of Vocal Foraging during Infant-Caregiver Communication," *Scientific Reports* 10, no. 1 (2020): 10469, https://doi.org/10.1038/s41598-020-66778-0.

31 **Studies of mothers:** Julie Gros-Louis et al., "Mothers Provide Differential Feedback to Infants' Prelinguistic Sounds," *International Journal of Behavioral Development* 30, no. 6 (November 2006): 509–16, https://doi.org/10.1177/0165025406071914.

32 **By five months old:** Michael H. Goldstein, Jennifer A. Schwade, and Marc H. Bornstein, "The Value of Vocalizing: Five-Month-Old Infants Associate Their Own Noncry Vocalizations with Responses from Caregivers," *Child Development* 80, no. 3 (2009): 636–44, https://doi.org/10.1111/j.1467-8624.2009.01287.x.

32 **For deaf infants:** Laura Ann Petitto and Paula F. Marentette, "Babbling in the Manual Mode: Evidence for the Ontogeny of Language," *Science* 251, no. 5000 (March 22, 1991): 1493–96, https://doi.org/10.1126/science.2006424.

32 **Little babblers are:** Ed Donnellan et al., "Infants' Intentionally Communicative Vocalizations Elicit Responses from Caregivers and Are the Best Predictors of the Transition to Language: A Longitudinal Investigation of Infants' Vocalizations, Gestures and Word Production," *Developmental Science* 23, no. 1 (2020): e12843, https://doi.org/10.1111/desc.12843.

33 **By babbling to an object:** Michael H. Goldstein et al., "Learning While Babbling: Prelinguistic Object-Directed Vocalizations Indicate a Readiness to Learn," *Infancy* 15, no. 4 (July–August 2010): 362–91, https://doi.org/10.1111/j.1532-7078.2009.00020.x.

34 **special ways of talking:** This article provides examples of variation in babytalk in other cultures: Clifton Pye, "Quiché Mayan Speech to Children," *Journal of Child Language* 13, no. 1 (February 1986): 85–100, https://doi.org/10.1017/S0305000900000313. This article analyzes properties of babytalk across many languages: M. Zettersten et al., "Evidence for Infant-Directed Speech Preference Is Consistent across Large-Scale, Multi-site Replication and Meta-Analysis," *Open Mind* 8 (2024): 439–61, https://doi.org/10.1162/opmi_a_00134.

34 **Whether spoken or signed:** Amanda S. Holzrichter and Richard P. Meier, "Child-Directed Signing in American Sign Language," in *Language Acquisition by Eye* (Lawrence Erlbaum, 2000), 25–40.

35 **In a recent survey:** Laura Wagner et al., "To What Extent Does the General Public Endorse Language Myths?," *Language and Linguistics Compass* 17, no. 3 (May/June 2023): e12486, https://doi.org/10.1111/lnc3.12486.

36 **You've undoubtedly heard:** This article both reviews some of the benefits of babytalk and describes similar speech directed to pets. Tobey Ben-Aderet et al., "Dog-Directed Speech: Why Do We Use It and Do Dogs Pay Attention to It?," *Proceedings of the Royal Society B: Biological Sciences* 284, no. 1846 (January 11, 2017): 20162429, https://doi.org/10.1098/rspb.2016.2429.

36 **The research about exaggerated intonation:** The first of these articles reports some downsides of babytalk, and the second reports some benefits: Andrew Martin et al., "Mothers Speak Less Clearly to Infants Than to Adults: A Comprehensive Test of the Hyperarticulation Hypothesis," *Psychological Science* 26, no. 3 (March 2015): 341–47, https://doi.org/10.1177

/0956797614562453. Erik D. Thiessen, Emily A. Hill, and Jenny R. Saffran, "Infant-Directed Speech Facilitates Word Segmentation," *Infancy* 7, no. 1 (2005): 53–71, https://doi.org/10.1207/s15327078in0701_5.

37 **Babytalk is the natural result:** Nicholas A. Smith and Laurel J. Trainor, "Infant-Directed Speech Is Modulated by Infant Feedback," *Infancy* 13, no. 4 (July–August 2008): 410–20, https://doi.org/10.1080/15250000802188719.

39 **We even know that infants:** Adrian Garcia-Sierra, Nairan Ramírez-Esparza, and Patricia K. Kuhl, "Relationships between Quantity of Language Input and Brain Responses in Bilingual and Monolingual Infants," *International Journal of Psychophysiology* 110 (December 2016): 1–17, https://doi.org/10.1016/j.ijpsycho.2016.10.004.

40 **Scientists think that for adults:** Jeremy I. Skipper, Howard C. Nusbaum, and Steven L. Small, "Listening to Talking Faces: Motor Cortical Activation during Speech Perception," *NeuroImage* 25, no. 1 (March 2005): 76–89, https://doi.org/10.1016/j.neuroimage.2004.11.006.

40 **The researchers studied six-month-old infants:** Alison G. Bruderer et al., "Sensorimotor Influences on Speech Perception in Infancy," *Proceedings of the National Academy of Sciences* 112, no. 44 (November 3, 2015): 13531–36, https://doi.org/10.1073/pnas.1508631112. At this writing, I don't know of any sign language versions of this phenomenon, so I'm describing hearing infants listening to language.

42 **We are better at this smile:** Adrienne Wood et al., "Fashioning the Face: Sensorimotor Simulation Contributes to Facial Expression Recognition," *Trends in Cognitive Sciences* 20, no. 3 (March 2016): 227–40, https://doi.org/10.1016/j.tics.2015.12.010.

43 **They found that toddler boys:** Paula M. Niedenthal et al., "Negative Relations between Pacifier Use and Emotional Competence," *Basic and Applied Social Psychology* 34, no. 5 (September 2012): 387–94, https://doi.org/10.1080/01973533.2012.712019. It's not clear why the effect didn't appear in little girls. Also, this study reflects correlations between long-term pacifier use and smile detection, rather than manipulations of pacifier use/not use, as would be analogous to the teether study.

43 **Some research suggests:** Alexandros K. Kanellopoulos and Sarah E. Costello, "The Effects of Prolonged Pacifier Use on Language Development in Infants and Toddlers," *Frontiers in Psychology* 15 (February 19, 2024), https://doi.org/10.3389/fpsyg.2024.1349323.

44 **And down the road:** Nadine Forget-Dubois et al., "Early Child Language Mediates the Relation between Home Environment and School Readiness," *Child Development* 80, no. 3 (May/June 2009): 736–49, https://doi.org/10.1111/j.1467-8624.2009.01294.x; Wayne A. Foster and Merideth Miller, "Development of the Literacy Achievement Gap: A Longitudinal Study of Kindergarten through Third Grade," *Language, Speech, and Hearing Services in Schools* 38, no. 3 (July 2007): 173–81, https://doi.org/10.1044/0161-1461(2007/018).

45 **Some (though not all) programs:** Jodi K. Heidlage et al., "The Effects of Parent-Implemented Language Interventions on Child Linguistic Outcomes: A Meta-Analysis," *Early Childhood Research Quarterly* 50, no. 1 (1st Quarter 2020): 6–23, https://doi.org/10.1016/j.ecresq.2018.12.006.

46 **amount of back-and-forth:** Rachel R. Romeo et al., "Beyond the 30-Million-Word Gap: Children's Conversational Exposure Is Associated with Language-Related Brain Function," *Psychological Science* 29, no. 5 (May 2018): 700–710, https://doi.org/10.1177/0956797617742725.

47 **Reading a book to a child:** Kate Nation, Nicola J. Dawson, and Yaling Hsiao, "Book Language and Its Implications for Children's Language, Literacy, and Development," *Current Directions in Psychological Science* 31, no. 4 (August 2022): 09637214221103264, https://doi.org/10.1177/09637214221103264; Jessica L. Montag, "Differences in Sentence Complexity in the Text of Children's Picture Books and Child-Directed Speech," *First Language* 39, no. 5 (October 2019): 527–46, https://doi.org/10.1177/0142723719849996.

47 **A toddler saying "kitty":** Evelina Fedorenko, Anna A. Ivanova, and Tamar I. Regev, "The Language Network as a Natural Kind within the Broader Landscape of the Human Brain," *Nature Reviews Neuroscience* 25, no. 5 (May 2024): 289–312, https://doi.org/10.1038/s41583-024-00802-4.

48 **Several programs designed:** Nicholas Dowdall et al., "Shared Picture Book

Reading Interventions for Child Language Development: A Systematic Review and Meta-Analysis," *Child Development* 91, no. 2 (March/April 2020): e383–99, https://doi.org/10.1111/cdev.13225.

48 **Alas, book reading with young:** This website shows current data from parent/caregiver surveys: "Increase the Proportion of Children Whose Family Read to Them at Least 4 Days per Week—EMC-02," Healthy People 2030, accessed January 8, 2024, https://odphp.health.gov/healthypeople/objectives-and-data/browse-objectives/children/increase-proportion-children-whose-family-read-them-least-4-days-week-emc-02.

49 **Parents who offer more media:** Suzy Tomopoulos et al., "Is Exposure to Media Intended for Preschool Children Associated with Less Parent-Child Shared Reading Aloud and Teaching Activities?," *Ambulatory Pediatrics* 7, no. 1 (January–February 2007): 18–24, https://doi.org/10.1016/j.ambp.2006.10.005.

49 **In 1970, children:** Yolanda (Linda) Reid Chassiakos et al., "Children and Adolescents and Digital Media," *Pediatrics* 138, no. 5 (November 2016): e20162593, https://doi.org/10.1542/peds.2016-2593. This report for the general public discusses rates of media use, pediatrician recommendations for lower media uses, the hype about media that is not actually beneficial for children, and many other topics.

50 ***Sesame Street*:** Shalom M. Fisch, Rosemarie T. Truglio, and Charlotte F. Cole, "The Impact of *Sesame Street* on Preschool Children: A Review and Synthesis of 30 Years' Research," *Media Psychology* 1, no. 2 (June 1999): 165–90, https://doi.org/10.1207/s1532785xmep0102_5.

51 **When kids are looking at screens:** Mary E. Brushe et al., "Screen Time and Parent-Child Talk When Children Are Aged 12 to 36 Months," *JAMA Pediatrics* 178, no. 4 (April 2024): 369–75, https://doi.org/10.1001/jamapediatrics.2023.6790.

51 **More screen time in preschoolers:** John S. Hutton et al., "Associations between Screen-Based Media Use and Brain White Matter Integrity in Preschool-Aged Children," *JAMA Pediatrics* 174, no. 1 (January 2020): e193869, https://doi.org/10.1001/jamapediatrics.2019.3869.

Chapter 3: The Challenge of Talking

53 **Sherlock Holmes:** Descriptions of Mycroft Holmes and the Diogenes Club are found in the story "The Greek Interpreter," in *The Memoirs of Sherlock Holmes* by Arthur Conan Doyle, available from Project Gutenberg: https://www.gutenberg.org/ebooks/834.

56 **Millions of people:** This website provides useful information about aphasia: "What Is Aphasia?" National Aphasia Association, accessed July 10, 2023, https://www.aphasia.org/aphasia-faqs/.

56 **Babies start to understand:** Elika Bergelson and Daniel Swingley, "At 6–9 Months, Human Infants Know the Meanings of Many Common Nouns," *Proceedings of the National Academy of Sciences* 109, no. 9 (February 28, 2012): 3253–58, https://doi.org/10.1073/pnas.1113380109.

57 **We go through our entire:** Clarence Green, "The Oral Language Productive Vocabulary Profile of Children Starting School: A Resource for Teachers," *Australian Journal of Education* 65, no. 1 (April 2021): 41–54, https://doi.org/10.1177/0004944120982771.

57 **That difference in practice:** Elyse K. Werner, "A Study of Communication Time" (master's thesis, University of Maryland, 1975).

58 **If speed-stacking:** "The Incredible Sport of Cup Stacking, Explained," posted January 25, 2017, by Vox, YouTube, https://www.youtube.com/watch?v=82DNYqurkxo.

58 **Similarly, speakers or singers:** Examples include a once-famous old FedEx commercial: "FedEx Commercial with John Moschitta," posted September 2, 2006, by ThreeOranges, YouTube, https://www.youtube.com/watch?v=NeK5ZjtpO-M; fast rapping: "Top 10 Fastest Rap Verses Ever," posted March 7, 2023, by WatchMojo.com, YouTube, https://www.youtube.com/watch?v=aNvVFncDo3s; and a chorus: "Knee Play 3 (live)—Philip Glass, 'Einstein on the Beach,'" posted October 31, 2012, by jpillow, YouTube, https://www.youtube.com/watch?v=PL9Rjn7EiRw.

59 **Per Ola Kristensson:** Heather Murphy, "Here's How to Type Faster on Your Phone," *New York Times*, October 4, 2019, https://www.nytimes.com/2019/10/04/technology/phone-typing.html.

60 **The brain works rapidly:** The classic book about talking, which worked out many of the issues described here, is Willem J. M. Levelt, *Speaking: From Intention to Articulation* (Bradford, 1989). More recent research can be found in Robert Hartsuiker and Kristof Strijkers, eds., *Language Production* (Routledge/Taylor & Francis, 2023).

61 **You know that feeling:** Lilla Magyari and Jan P. de Ruiter, "Prediction of Turn-Ends Based on Anticipation of Upcoming Words," *Frontiers in Psychology* 3 (October 2012): 376, http://journal.frontiersin.org/article/10.3389/fpsyg.2012.00376/full.

61 **Podcast developers:** Jane Hu, "What's Your 'X' Rating?," *Audible* (blog), June 8, 2017, https://www.audible.com/blog/the-listening-life/whats-your-x-rating/.

61 **That result corresponds:** There is wide variation in estimates of speaking rates and listening (and reading) rates in English. Adults probably can comprehend up to about two hundred to three hundred spoken or written English words per minute, depending on the topic and many other factors. Speaking rates in English roughly average one hundred to one hundred and fifty words per minute, again varying with many factors. Handwriting or typing on a keyboard or smartphone are all slower. Victor Kuperman et al., "A Lingering Question Addressed: Reading Rate and Most Efficient Listening Rate Are Highly Similar," *Journal of Experimental Psychology: Human Perception and Performance* 47, no. 8 (2021): 1103–12, https://doi.org/10.1037/xhp0000932; Howard Maclay and Charles E. Osgood, "Hesitation Phenomena in Spontaneous English Speech," *WORD* 15, no. 1 (January 1959): 19–44, https://doi.org/10.1080/00437956.1959.11659682.

63 **Amit's research team:** Timothy W. Boiteau et al., "Interference between Conversation and a Concurrent Visuomotor Task," *Journal of Experimental Psychology: General* 143, no. 1 (February 2014): 295–311, https://doi.org/10.1037/a0031858.

64 **Now suddenly the listener:** Magyari and de Ruiter, "Prediction of Turn-Ends."

64 **Indeed, devising the plan:** Susan Kemper et al., "Tracking Talking: Dual Task Costs of Planning and Producing Speech for Young versus Older

Adults," *Aging, Neuropsychology, and Cognition* 18, no. 3 (May 2011): 257–79, https://doi.org/10.1080/13825585.2010.527317.

64 **Studies using a driving simulator:** Ensar Becic et al., "Driving Impairs Talking," *Psychonomic Bulletin & Review* 17, no. 1 (February 2010): 15–21, https://doi.org/10.3758/PBR.17.1.15.

65 **Hundreds of studies using fMRI:** Evelina Fedorenko, Anna A. Ivanova, and Tamar I. Regev, "The Language Network as a Natural Kind within the Broader Landscape of the Human Brain," *Nature Reviews Neuroscience* 25, no. 5 (May 2024): 289–312, https://doi.org/10.1038/s41583-024-00802-4.

65 **A second discovery:** Gina F. Humphreys and Silvia P. Gennari, "Competitive Mechanisms in Sentence Processing: Common and Distinct Production and Reading Comprehension Networks Linked to the Prefrontal Cortex," *NeuroImage* 84 (January 2014): 354–66, https://www.sciencedirect.com/science/article/abs/pii/S1053811913009233.

66 **Together, these shortcuts yield:** Mark J. Koranda, Martin Zettersten, and Maryellen C. MacDonald, "Good-Enough Production: Selecting Easier Words Instead of More Accurate Ones," *Psychological Science* 33, no. 9 (September 2022): 1440–51, https://doi.org/10.1177/09567976221089603.

67 **Zadie Smith's novel:** Zadie Smith, *On Beauty* (Thorndike Press, 2006).

67 **This supplementary gesture:** The gesture expert Martha Alibali mentioned this example to me years ago, and it's one of my favorites. Martha W. Alibali et al., "Gesture-Speech Integration in Narrative: Are Children Less Redundant Than Adults?," *Gesture* 9, no. 3 (January 2009): 290–311, https://doi.org/10.1075/gest.9.3.02ali.

68 **This helpful feedback:** Birgit Knudsen, Ava Creemers, and Antje S. Meyer, "Forgotten Little Words: How Backchannels and Particles May Facilitate Speech Planning in Conversation?," *Frontiers in Psychology* 11 (November 2020): 593671, https://doi.org/10.3389/fpsyg.2020.593671.

69 **Talking is clearly not:** Morten H. Christiansen and Nick Chater, "The Now-or-Never Bottleneck: A Fundamental Constraint on Language," *Be-*

havioral and Brain Sciences 39 (2016): e62, https://doi.org/10.1017/S0140525X1500031X.

70 **Starting sentences with common:** Maryellen C. MacDonald, "How Language Production Shapes Language Form and Comprehension," *Frontiers in Psychology* 4 (April 2013): 226, https://doi.org/10.3389/fpsyg.2013.00226.

71 **In both spoken and signed languages:** Stephen C. Levinson, "Turn-Taking in Human Communication—Origins and Implications for Language Processing," *Trends in Cognitive Sciences* 20, no. 1 (January 2016): 6–14, https://doi.org/10.1016/j.tics.2015.10.010.

71 **A long gap signals:** Emma M. Templeton et al., "Long Gaps between Turns Are Awkward for Strangers but Not for Friends," *Philosophical Transactions of the Royal Society B: Biological Sciences* 378, no. 1875 (March 6, 2023): 20210471, https://doi.org/10.1098/rstb.2021.0471.

71 **Conversational turn taking is said:** Levinson, "Turn-Taking in Human Communication."

73 **But neither humans:** Brett M. Gibson, Edward A. Wasserman, and Alan C. Kamil, "Pigeons and People Select Efficient Routes When Solving a One-Way 'Traveling Salesperson' Task," *Journal of Experimental Psychology: Animal Behavior Processes* 33, no. 3 (July 2007): 244–61, https://doi.org/10.1037/0097-7403.33.3.244; Hiromitsu Miyata and Kazuo Fujita, "Flexible Route Selection by Pigeons (*Columba livia*) on a Computerized Multi-Goal Navigation Task with and without an 'Obstacle,'" *Journal of Comparative Psychology* 125, no. 4 (November 2011): 431–35, https://doi.org/10.1037/a0024240.

73 **Musical improvisation is complex:** Roger E. Beaty et al., "Expert Musical Improvisations Contain Sequencing Biases Seen in Language Production," *Journal of Experimental Psychology: General* 151, no. 4 (2021): 912–20, https://doi.org/10.1037/xge0001107.

74 **Studies of toothbrushing:** I. D. M. Macgregor and A. J. Rugg-Gunn, "A Survey of Toothbrushing Sequence in Children and Young Adults," *Journal of Periodontal Research* 14, no. 3 (June 1979): 225–30, https://doi.org/10.1111/j.1600-0765.1979.tb00227.x.

Chapter 4: Talking and Mental Focus

79 **"The ancients etched":** "The Sadness of Clothes" by Emily Fragos contains other wonderful lines about talking besides the ones quoted here. The poem can be found at https://poets.org/poem/sadness-clothes, and you can listen to a beautiful reading of it at https://theamericanscholar.org/the-sadness-of-clothes-by-emily-fragos/.

79 **About three thousand years ago:** "Zarathustra," World History Encyclopedia, May 28, 2020, https://www.worldhistory.org/zoroaster/.

80 **Child-development experts believe:** Jane S. M. Lidstone, Elizabeth Meins, and Charles Fernyhough, "The Roles of Private Speech and Inner Speech in Planning during Middle Childhood: Evidence from a Dual Task Paradigm," *Journal of Experimental Child Psychology* 107, no. 4 (December 2010): 438–51, https://doi.org/10.1016/j.jecp.2010.06.002.

81 **Internal self-talk is nonetheless:** If you'd like to dive deeper into self-talk, a good start is this book: Ethan Kross, *Chatter: The Voice in Our Head, Why It Matters, and How to Harness It* (Penguin Random House, 2021).

81 **We can think of it:** An alternative view is here: Evelina Fedorenko, Steven T. Piantadosi, and Edward A. F. Gibson, "Language Is Primarily a Tool for Communication Rather Than Thought," *Nature* 630, no. 8017 (June 2024): 575–86, https://doi.org/10.1038/s41586-024-07522-w.

81 **Sensitive equipment can:** Emily Matchar, "This Device Can Hear You Talking to Yourself," *Smithsonian Magazine*, August 5, 2019, https://www.smithsonianmag.com/innovation/device-can-hear-voice-inside-your-head-180972785/.

81 **And sophisticated computer programs:** Pam Belluck, "A Stroke Stole Her Ability to Speak at 30. A.I. Is Helping to Restore It Years Later," *New York Times*, August 23, 2023, https://www.nytimes.com/2023/08/23/health/ai-stroke-speech-neuroscience.html.

81 **Interestingly, people vary:** Johanne S. K. Nedergaard and Gary Lupyan, "Not Everybody Has an Inner Voice: Behavioral Consequences of Anendophasia," *Psychological Science* 35, no. 7 (July 2024): 780–97, https://doi.org/10.1177/09567976241243004.

82 **Researchers initially thought of:** Charles Fernyhough and Anna M. Borghi, "Inner Speech as Language Process and Cognitive Tool," *Trends in Cognitive Sciences* 27, no. 12 (December 2023): 1180–93, https://doi.org/10.1016/j.tics.2023.08.014.

83 **These skills aren't just developing in parallel:** Laura J. Kuhn et al., "The Contribution of Children's Time-Specific and Longitudinal Expressive Language Skills on Developmental Trajectories of Executive Function," *Journal of Experimental Child Psychology* 148 (August 2016): 20–34, https://doi.org/10.1016/j.jecp.2016.03.008; Pauline L. Slot and Antje von Suchodoletz, "Bidirectionality in Preschool Children's Executive Functions and Language Skills: Is One Developing Skill the Better Predictor of the Other?," *Early Childhood Research Quarterly* 42 (1st Quarter 2018): 205–14, https://doi.org/10.1016/j.ecresq.2017.10.005.

83 **Kids who engage in more conversations:** Amy E. Carolus et al., "Conversation Disruptions in Early Childhood Predict Executive Functioning Development: A Longitudinal Study," *Developmental Science* 27, no. 1 (January 2024): e13414, https://doi.org/10.1111/desc.13414.

85 **People who are searching:** Gary Lupyan and Daniel Swingley, "Self-Directed Speech Affects Visual Search Performance," *Quarterly Journal of Experimental Psychology* 65, no. 6 (June 2012): 1068–85, https://doi.org/10.1080/17470218.2011.647039.

85 **Five-year-olds who are trying:** Sabine Doebel et al., "Using Language to Get Ready: Familiar Labels Help Children Engage Proactive Control," *Journal of Experimental Child Psychology* 166 (February 2018): 147–59, https://doi.org/10.1016/j.jecp.2017.08.006.

85 **Adults who are trying to discover:** Martin Zettersten and Gary Lupyan, "Finding Categories through Words: More Nameable Features Improve Category Learning," *Cognition* 196 (March 2020): 104135, https://doi.org/10.1016/j.cognition.2019.104135.

89 **This phenomenon is called verbal overshadowing:** Jonathan W. Schooler and Tonya Y. Engstler-Schooler, "Verbal Overshadowing of Visual Memories: Some Things Are Better Left Unsaid," *Cognitive Psychology* 22, no. 1 (January 1990): 36–71, https://doi.org/10.1016/0010-0285(90)90003-M.

90 **Interestingly, verbal overshadowing doesn't even:** Mark J. Koranda and Maryellen C. MacDonald, "Language and Gesture Descriptions Affect Memory: A Nonverbal Overshadowing Effect" in *Proceedings of the 37th Annual Meeting of the Cognitive Science Society*, 2015, https://www.semanticscholar.org/paper/Language-and-Gesture-Descriptions-Affect-Memory-A-Koranda-MacDonald/985b8503daf9245659467c49c077c64adbbd3485.

90 **Drawing a picture:** Wilma A. Bainbridge et al., "Drawing as a Means to Characterize Memory and Cognition," *Memory & Cognition* (August 2024), https://doi.org/10.3758/s13421-024-01618-4.

91 **In this study, people who were:** Katharina Kircanski, Matthew D. Lieberman, and Michelle G. Craske, "Feelings into Words: Contributions of Language to Exposure Therapy," *Psychological Science* 23, no. 10 (October 2012): 1086–91, https://doi.org/10.1177/0956797612443830.

92 **Brain-imaging methods suggest:** Matthew D. Lieberman et al., "Putting Feelings into Words," *Psychological Science* 18, no. 5 (May 2007): 421–28, https://doi.org/10.1111/j.1467-9280.2007.01916.x.

93 **One explanation for this foreign-language:** Boaz Keysar, Sayuri L. Hayakawa, and Sun Gyu An, "The Foreign-Language Effect: Thinking in a Foreign Tongue Reduces Decision Biases," *Psychological Science* 23, no. 6 (June 2012): 661–68, https://doi.org/10.1177/0956797611432178; Nicola Del Maschio et al., "Decision-Making Depends on Language: A Meta-Analysis of the Foreign Language Effect," *Bilingualism: Language and Cognition* 25, no. 4 (August 2022): 617–30, https://doi.org/10.1017/S1366728921001012.

94 **Hellen Obiri is an elite:** "Chebet Retains Boston Marathon Men's Title, Obiri Wins Women's Race," *Fiji Broadcasting Corporation* (blog), April 18, 2023, https://www.fbcnews.com.fj/sports/athletics/chebet-retains-boston-marathon-mens-title-obiri-wins-womens-race/.

95 **The snowboarder Shaun White:** White's *who cares* and other examples of self-talk are in this YouTube video, including Muhammad Ali's self-talk before the Ali–Foreman fight: "Self-Talk—How Your Favourite Athletes LEVEL UP Their Mental Game," posted February 23, 2023, by Sam Martin—Peak Performance, YouTube, https://www.youtube.com/watch?v=CdLbRgWiC00.

96 **To aid any exploration:** This review of self-talk in sports mentions relevant findings for the recommendations here. Judy L. Van Raalte, Andrew Vincent, and Britton W. Brewer, "Self-Talk: Review and Sport-Specific Model," *Psychology of Sport and Exercise* 22 (January 2016): 139–48, https://doi.org/10.1016/j.psychsport.2015.08.004.

97 **Michael Cunningham's novel:** Michael Cunningham, *Flesh and Blood* (Picador, 2007).

98 **One example is a gratitude journal:** Lilian Jans-Beken et al., "Gratitude and Health: An Updated Review," *Journal of Positive Psychology* 15, no. 6 (November 2020): 743–82, https://doi.org/10.1080/17439760.2019.1651888.

98 **The endnotes have some suggestions:** Here's a short article reviewing the benefits of keeping a journal along with some tips for getting started: Margarita Tartakovsky, "6 Journaling Benefits and How to Start Right Now," Healthline, February 22, 2022, https://www.healthline.com/health/benefits-of-journaling#how-to-start.

99 **Studies of prayer:** Sarah A. Schnitker and Kelsy L. Richardson, "Framing Gratitude Journaling as Prayer Amplifies Its Hedonic and Eudaimonic Well-Being, but Not Health, Benefits," *Journal of Positive Psychology* 14, no. 4 (July 2019): 427–39, https://doi.org/10.1080/17439760.2018.1460690.

100 **The writing exercises continue:** This is Jamie Pennebaker's guide to expressive writing, which reviews research results and is full of suggestions for how to do it: James W. Pennebaker and Joshua M. Smyth, *Opening Up by Writing It Down: How Expressive Writing Improves Health and Eases Emotional Pain*, 3rd ed. (Guilford Press, 2016).

101 **"I write," said the prolific essayist:** Joan Didion, "Why I Write," *New York Times*, December 5, 1976, https://www.nytimes.com/1976/12/05/archives/why-i-write-why-i-write.html.

102 **Judy and her collaborators:** Chris S. Hulleman and Judith M. Harackiewicz, "Promoting Interest and Performance in High School Science Classes," *Science* 326, no. 5958 (December 4, 2009): 1410–12, https://doi.org/10.1126/science.1177067; Michael W. Asher et al., "Utility-Value Intervention Promotes Persistence and Diversity in STEM," *Proceedings of the National*

Academy of Sciences 120, no. 19 (May 1, 2023): e2300463120, https://doi.org/10.1073/pnas.2300463120.

106 **Rumination is often described:** This review defines rumination, reviews its negative consequences, and describes attempts to improve mental health, including targeting rumination itself: Edward R. Watkins and Henrietta Roberts, "Reflecting on Rumination: Consequences, Causes, Mechanisms and Treatment of Rumination," *Behaviour Research and Therapy* 127 (April 2020): 103573, https://doi.org/10.1016/j.brat.2020.103573.

106 **Rumination is typically self-talk:** Amanda J. Rose, "The Costs and Benefits of Co-Rumination," *Child Development Perspectives* 15, no. 3 (September 2021): 176–81, https://doi.org/10.1111/cdep.12419.

106 **Like other self-talk, rumination:** Vera Vine, Amelia Aldao, and Susan Nolen-Hoeksema, "Chasing Clarity: Rumination as a Strategy for Making Sense of Emotions," *Journal of Experimental Psychopathology* 5, no. 3 (November 2014): 229–43, https://doi.org/10.5127/jep.038513.

107 **There are some encouraging results:** Watkins and Roberts, "Reflecting on Rumination."

108 **Aza's internal OCD monologue:** John Green, *Turtles All the Way Down* (Penguin Random House, 2019), 46–47.

108 **Mindfulness encourages practitioners:** Jean L. Kristeller, "Mindfulness Meditation," in *Principles and Practice of Stress Management*, 3rd ed., ed. Paul M. Lehrer, Robert L. Woolfolk, and Wesley E. Sime (Guilford Press, 2007).

109 **We've seen his triangle:** R. C. Zaehner, *The Teachings of the Magi: A Compendium of Zoroastrian Beliefs* (Routledge, 2021), https://doi.org/10.4324/9781003241461.

109 **Hate speech is the topic:** Here is a sample of articles on different aspects of the problem of hate speech: María Antonia Paz, Julio Montero-Díaz, and Alicia Moreno-Delgado, "Hate Speech: A Systematized Review," *Sage Open* 10, no. 4 (October–December 2020), https://doi.org/10.1177/2158244020973022; Binny Mathew et al., "Hate Begets Hate: A Temporal Study of Hate Speech," *Proceedings of the ACM on Human-Computer Inter-*

action 4, no. CSCW2 (October 2020): 1–24, https://doi.org/10.1145/3415163; Carmen Cervone, Martha Augoustinos, and Anne Maass, "The Language of Derogation and Hate: Functions, Consequences, and Reappropriation," *Journal of Language and Social Psychology* 40, no. 1 (January 2021): 80–101, https://doi.org/10.1177/0261927X20967394; Matthew L. Williams et al., "Hate in the Machine: Anti-Black and Anti-Muslim Social Media Posts as Predictors of Offline Racially and Religiously Aggravated Crime," *British Journal of Criminology* 60, no. 1 (January 2020): 93–117, https://doi.org/10.1093/bjc/azz049.

110 **Producing hate speech amplifies the belief:** James W. Downing, Charles M. Judd, and Markus Brauer, "Effects of Repeated Expressions on Attitude Extremity," *Journal of Personality and Social Psychology* 63, no. 1 (1992): 17–29, https://doi.org/10.1037/0022-3514.63.1.17; Markus Brauer and Charles M. Judd, "Group Polarization and Repeated Attitude Expressions: A New Take on an Old Topic," *European Review of Social Psychology* 7, no. 1 (January 1996): 173–207, https://doi.org/10.1080/14792779643000010.

110 **Talking about something makes it:** Melissa C. Duff and Sarah Brown-Schmidt, "The Hippocampus and the Flexible Use and Processing of Language," *Frontiers in Human Neuroscience* 6 (2012): 69, https://doi.org/10.3389/fnhum.2012.00069.

110 **More mentions breed more familiarity:** Ullrich K. H. Ecker et al., "The Psychological Drivers of Misinformation Belief and Its Resistance to Correction," *Nature Reviews Psychology* 1, no. 1 (January 2022): 13–29, https://doi.org/10.1038/s44159-021-00006-y.

Chapter 5: Talking to Learn

113 **"The hippocampus is the part":** Tommy Orange, *There There* (Penguin Random House, 2019).

116 **they compared the standard college:** Peter Reuell, "Study Shows Students in 'Active Learning' Classrooms Learn More Than They Think," *Harvard Gazette*, September 4, 2019, https://news.harvard.edu/gazette/story/2019/09/study-shows-that-students-learn-more-when-taking-part-in-classrooms-that-employ-active-learning-strategies/.

116 **Active learning can:** A good magazine article about active learning is here: Aatish Bhatia, "Active Learning Leads to Higher Grades and Fewer Failing Students in Science, Math, and Engineering," *Wired*, May 12, 2014, https://www.wired.com/2014/05/empzeal-active-learning/.

117 **The physics researchers put it:** The ease-of-listening quote comes from the Deslauriers et al. section of this multi-section article: Nesra Yannier et al., "Active Learning: 'Hands-On' Meets 'Minds-On,'" *Science* 374, no. 6563 (October 2021): 26–30, https://doi.org/10.1126/science.abj9957.

118 **Learning happens gradually:** James J. Knierim, "The Hippocampus," *Current Biology* 25, no. 23 (December 7, 2015): R1116–21, https://doi.org/10.1016/j.cub.2015.10.049.

119 **Making choices is part:** Michael J. Carter and Diane M. Ste-Marie, "Not All Choices Are Created Equal: Task-Relevant Choices Enhance Motor Learning Compared to Task-Irrelevant Choices," *Psychonomic Bulletin & Review* 24, no. 6 (December 2017): 1879–88, https://doi.org/10.3758/s13423-017-1250-7.

119 **The 2022 scores:** You can look at scores for the 2022 NAEP (pronounced "nape," often known as the nation's report card) for US fourth and eighth graders here, including breakdowns by race, region, and other groupings. The eighth-grade results show similar declines to the fourth-grade results described in the text. "Scores Decline in NAEP Reading at Grades 4 and 8 Compared to 2019," Report Card: 2022 NAEP Reading Assessment, accessed September 27, 2024, https://www.nationsreportcard.gov/highlights/reading/2022/. There are also NAEP scores for math and some other parts of the elementary-secondary school curriculum in the United States.

122 **Reading aloud focuses the child's attention:** Cathy L. Fiala and Susan M. Sheridan, "Parent Involvement and Reading: Using Curriculum-Based Measurement to Assess the Effects of Paired Reading," *Psychology in the Schools* 40, no. 6 (2003): 613–26, https://doi.org/10.1002/pits.10128.

122 **The PBS show:** US Department of Education, "25 Activities for Reading and Writing Fun," Reading Rockets, accessed July 17, 2023, https://www.readingrockets.org/topics/activities/articles/25-activities-reading-and-writing-fun.

123 **Children who have poorer spoken-language skills:** Margaret J. Snowling and Charles Hulme, "Annual Research Review: Reading Disorders Revisited—the Critical Importance of Oral Language," *Journal of Child Psychology and Psychiatry* 62, no. 5 (May 2021): 635–53, https://doi.org/10.1111/jcpp.13324.

123 **Written language, even in books:** Kate Nation, Nicola J. Dawson, and Yaling Hsiao, "Book Language and Its Implications for Children's Language, Literacy, and Development," *Current Directions in Psychological Science* 34, no. 4 (August 2022): 375–80, https://doi.org/10.1177/09637214221103264.

124 **The NELI program:** This link, https://oxedandassessment.com/uk/neli/, goes to the NELI program's home page, and on that page are additional links to videos, further detail about the program, and a list of published research reports on NELI's effectiveness. This article reviews both NELI and other language interventions and discusses the importance of these programs for young children's education: Charles Hulme et al., "Children's Language Skills Can Be Improved: Lessons from Psychological Science for Educational Policy," *Current Directions in Psychological Science* 29, no. 4 (August 2020): 372–77, https://doi.org/10.1177/0963721420923684.

125 **The comprehensible-input idea is that:** Stephen Krashen, "The Case for Comprehensible Input," *Language Magazine*, July 17, 2017, https://www.languagemagazine.com/2017/07/17/case-for-comprehension/.

126 **A graduate student and I directly investigated:** Elise W. M. Hopman and Maryellen C. MacDonald, "Producing during Language Learning Improves Comprehension," *Psychological Science* 29, no. 6 (2018): 961–71, https://doi.org/10.1177/0956797618754486.

127 **Other studies of the benefits:** Valérie Keppenne, Elise W. M. Hopman, and Carrie N. Jackson, "Production-Based Training Benefits the Comprehension and Production of Grammatical Gender in L2 German," *Applied Psycholinguistics* 42, no. 4 (July 2021): 907–36, https://doi.org/10.1017/S014271642100014X; Elisabeth Wilhelmina Maria Hopman, "Modality Matters: Generalization in Second Language Learning after Production versus Comprehension Practice" (PhD diss., University of Wisconsin–Madison, 2022), ProQuest (29255039), https://www.proquest.com/pqdtglobal/docview/2687737040/abstract/B214268077294B8EPQ/3.

127 **Indeed, brain imaging studies show:** Diana López-Barroso et al., "Word Learning Is Mediated by the Left Arcuate Fasciculus," *Proceedings of the National Academy of Sciences* 110, no. 32 (August 6, 2013): 13168–73, https://doi.org/10.1073/pnas.1301696110.

129 **A few attempts to teach children:** This book discusses a number of talk-based projects for young students, with varying outcomes. The introduction provides a good overview, and chapters 8 and 10 show some particularly long-lasting benefits. Lauren B. Resnick, Christa S. C. Asterhan, and Sherice N. Clarke, *Socializing Intelligence Through Academic Talk and Dialogue* (American Educational Research Association, 2015), https://www.jstor.org/stable/j.ctt1s474m1.

129 **Two reading researchers:** The reading researchers who visited this no-talking school, and who both told me about it later, were Julie Washington and Mark Seidenberg.

130 **What is well documented:** Daniel J. Losen et al., "Are We Closing the School Discipline Gap?," Civil Rights Project, February 23, 2015, https://escholarship.org/uc/item/2t36g571.

130 **And active-learning activities foster:** Elli J. Theobald et al., "Active Learning Narrows Achievement Gaps for Underrepresented Students in Undergraduate Science, Technology, Engineering, and Math," *Proceedings of the National Academy of Sciences* 117, no. 12 (March 24, 2020): 6476–83, https://doi.org/10.1073/pnas.1916903117.

131 **Questions can be incorporated:** Kara Kedrick, Paul Schrater, and Wilma Koutstaal, "The Multifaceted Role of Self-Generated Question Asking in Curiosity-Driven Learning," *Cognitive Science* 47, no. 4 (April 2023): e13253, https://doi.org/10.1111/cogs.13253; the specific example of children imagining asking an author questions is in Resnick, Asterhan, and Clarke, *Socializing Intelligence*.

131 **When children and adults generate explanations:** Erik Brockbank et al., "Ask Me Why, Don't Tell Me Why: Asking Children for Explanations Facilitates Relational Thinking," *Developmental Science* 26, no. 1 (January 2023): e13274, https://doi.org/10.1111/desc.13274; Tania Lombrozo and Emily G. Liquin, "Explanation Is Effective Because It Is Selective," *Current*

Directions in Psychological Science 32, no. 3 (June 2023): 212–19, https://doi.org/10.1177/09637214231156106.

132 **We remember things we say ourselves:** Sarah Brown-Schmidt and Aaron S. Benjamin, "How We Remember Conversation: Implications in Legal Settings," *Policy Insights from the Behavioral and Brain Sciences* 5, no. 2 (October 2018): 187–94, https://doi.org/10.1177/2372732218786975.

132 **The act of recalling information:** Jeffrey D. Karpicke, "Retrieval-Based Learning: Active Retrieval Promotes Meaningful Learning," *Current Directions in Psychological Science* 21, no. 3 (June 2012): 157–63, https://doi.org/10.1177/0963721412443552.

132 **Generating an explanation improves learning:** Pooja G. Sidney, Shanta Hattikudur, and Martha W. Alibali, "How Do Contrasting Cases and Self-Explanation Promote Learning? Evidence from Fraction Division," *Learning and Instruction* 40 (December 2015): 29–38, https://doi.org/10.1016/j.learninstruc.2015.07.006.

134 **gesturing and drawing:** Judith E. Fan, "Drawing to Learn: How Producing Graphical Representations Enhances Scientific Thinking," *Translational Issues in Psychological Science* 1, no. 2 (June 2015): 170–81, https://doi.org/10.1037/tps0000037; Susan Goldin-Meadow, "How Gesture Promotes Learning throughout Childhood," *Child Development Perspectives* 3, no. 2 (August 2009): 106–11, https://doi.org/10.1111/j.1750-8606.2009.00088.x.

135 **Second, tests are learning events:** Henry L. Roediger and Jeffrey D. Karpicke, "Test-Enhanced Learning: Taking Memory Tests Improves Long-Term Retention," *Psychological Science* 17, no. 3 (March 2006): 249–55, https://doi.org/10.1111/j.1467-9280.2006.01693.x.

136 **The researchers who studied those physics students:** Louis Deslauriers et al., "Measuring Actual Learning versus Feeling of Learning in Response to Being Actively Engaged in the Classroom," *Proceedings of the National Academy of Sciences* 116, no. 39 (September 24, 2019): 19251–57, https://doi.org/10.1073/pnas.1821936116.

137 ***The Wall Street Journal* reports:** This link reports earnings of various college majors. "Salary Increase by Major," *Wall Street Journal*, accessed January 8, 2024, https://www.wsj.com/public/resources/documents/info

-Degrees_that_Pay_you_Back-sort.html. This link reports test scores on graduate exams (GRE, LSAT, etc.) for various college majors: "Value of Philosophy—Charts and Graphs," Daily Nous, accessed June 12, 2024, https://dailynous.com/value-of-philosophy/charts-and-graphs/.

137 **If you've ever seen movies:** "Examples of Socratic Method in Law School," JD Advising, accessed February 18, 2024, https://jdadvising.com/examples-of-the-socratic-method-in-law-school/.

138 **The Discussion Project is the brainchild:** This website is the home page of the project: https://discussion.education.wisc.edu/. It also has additional links to examples and detail about the curriculum.

139 **The Semester in Dialogue:** This page contains information and additional links for more in-depth information: "About," Morris J. Wosk Centre for Dialogue, Simon Fraser University, accessed February 19, 2022, https://sfu.ca/dialogue/learn/semester-in-dialogue/about.html.

Chapter 6: Talking and Aging Well

143 **In Felix Salten's 1923 novel *Bambi*:** Felix Salten, *Bambi: A Life in the Woods*, trans. Whittaker Chambers (Simon & Schuster, 1928). The leaf story is in chapter 8 and the book is available from Project Gutenberg, https://www.gutenberg.org/ebooks/72577. The translator, Whittaker Chambers, is the same Whittaker Chambers who was a Soviet spy in the United States in the 1930s and who later defected and cooperated with the US government, testifying against Alger Hiss and others.

145 **The one that older adults worry:** This report describes data from a national survey of adults thirty and older, identifying their concerns and opinions in each decade: "Perceptions of Aging during Each Decade of Life after 30," West Health Institute, NORC at the University of Chicago, March 2017, https://www.norc.org/PDFs/WHI-NORC-Aging-Survey/Brief_WestHealth_A_2017-03_DTPv2.pdf.

145 **Because we're tending to live longer:** The Alzheimer's Association has a useful document about many aspects of dementia, including the ones discussed here, "Alzheimer's Disease Facts and Figures," Alzheimer's Associa-

tion, accessed July 16, 2023, https://www.alz.org/alzheimers-dementia/facts-figures. Another publication that discusses rates of cognitive impairment and dementia in the United States is here: "Fact Sheet: U.S. Dementia Trends," PRB, October 21, 2021, https://www.prb.org/resources/fact-sheet-u-s-dementia-trends/.

146 **Researchers have uncovered about a dozen:** Gill Livingston et al., "Dementia Prevention, Intervention, and Care: 2020 Report of the *Lancet* Commission," *The Lancet* 396, no. 10248 (August 8, 2020): 413–46, https://doi.org/10.1016/S0140-6736(20)30367-6.

147 **Vocabulary is a major:** Gitit Kavé, "Vocabulary Changes in Adulthood: Main Findings and Methodological Considerations," *International Journal of Language & Communication Disorders* 59, no. 1 (January/February 2024): 58–67, https://doi.org/10.1111/1460-6984.12820.

147 **Word-finding problems increase:** Meredith A. Shafto et al., "Age-Related Increases in Verbal Knowledge Are Not Associated with Word Finding Problems in the Cam-CAN Cohort: What You Know Won't Hurt You," *Journals of Gerontology: Series B* 72, no. 1 (January 1, 2017): 100–106, https://doi.org/10.1093/geronb/gbw074.

148 **There is some evidence linking vocabulary:** Michael Ramscar et al., "The Myth of Cognitive Decline: Non-Linear Dynamics of Lifelong Learning," *Topics in Cognitive Science* 6, no. 1 (January 2014): 5–42, https://doi.org/10.1111/tops.12078.

148 **For example, it takes extra effort:** Xiaorong Cheng, Graham Schafer, and Elkan G. Akyürek, "Name Agreement in Picture Naming: An ERP Study," *International Journal of Psychophysiology* 76, no. 3 (June 2010): 130–41, https://doi.org/10.1016/j.ijpsycho.2010.03.003.

149 **We can see similar effects in bilinguals:** Tamar H. Gollan and Lori-Ann R. Acenas, "What Is a TOT? Cognate and Translation Effects on Tip-of-the-Tongue States in Spanish-English and Tagalog-English Bilinguals," *Journal of Experimental Psychology: Learning, Memory, and Cognition* 30, no. 1 (January 2004): 246–69, https://doi.org/10.1037/0278-7393.30.1.246.

149 **These abilities also start sliding:** Heather J. Ferguson, Victoria E. A. Brunsdon, and Elisabeth E. F. Bradford, "The Developmental Trajectories of

Executive Function from Adolescence to Old Age," *Scientific Reports* 11, no. 1 (January 14, 2021): 1382, https://doi.org/10.1038/s41598-020-80866-1.

149 **For example, adults with ADHD:** Paul E. Engelhardt, Fernanda Ferreira, and Joel T. Nigg, "Language Production Strategies and Disfluencies in Multi-Clause Network Descriptions: A Study of Adult Attention-Deficit/Hyperactivity Disorder," *Neuropsychology* 25, no. 4 (July 2011): 442–53, https://doi.org/10.1037/a0022436.

150 **They substitute easy words:** Meredith A. Shafto and Lorraine K. Tyler, "Language in the Aging Brain: The Network Dynamics of Cognitive Decline and Preservation," *Science* 346, no. 6209 (October 31, 2014): 583–87, https://doi.org/10.1126/science.1254404.

152 **There are a few medications:** This website offers a list of currently available medications to treat memory impairments: "Medications for Memory, Cognition and Dementia-Related Behaviors," Alzheimer's Association, accessed August 22, 2023, https://www.alz.org/alzheimers-dementia/treatments/medications-for-memory. And this website discusses over-the-counter supplements for memory: Mayo Clinic Staff, "Ginkgo," Mayo Clinic, accessed August 22, 2023, https://www.mayoclinic.org/drugs-supplements-ginkgo/art-20362032.

152 **Those with cognitive reserves:** Monica E. Nelson et al., "Cognitive Reserve, Alzheimer's Neuropathology, and Risk of Dementia: A Systematic Review and Meta-Analysis," *Neuropsychology Review* 31, no. 2 (June 2021): 233–50, https://doi.org/10.1007/s11065-021-09478-4.

152 **One way to build up:** Additional discussion of the effects of bilingualism on cognition can be found in this book: Viorica Marian, *The Power of Language: How the Codes We Use to Think, Speak, and Live Transform Our Minds* (Dutton, 2023).

152 **Bilinguals who use different languages:** Merve Gul Degirmenci et al., "The Role of Bilingualism in Executive Functions in Healthy Older Adults: A Systematic Review," *International Journal of Bilingualism* 26, no. 4 (August 2022): 426–49, https://doi.org/10.1177/13670069211051291; Emily S. Nichols et al., "Bilingualism Affords No General Cognitive Advantages: A Population Study of Executive Function in 11,000 People,"

Psychological Science 31, no. 5 (May 2020): 548–67, https://doi.org/10.1177/0956797620903113.

154 **Many investigations of these games:** Monica Melby-Lervåg and Charles Hulme, "Is Working Memory Training Effective? A Meta-Analytic Review," *Developmental Psychology* 49, no. 2 (February 2013): 270–91, https://doi.org/10.1037/a0028228; Luc Watrin, Gizem Hülür, and Oliver Wilhelm, "Training Working Memory for Two Years—No Evidence of Transfer to Intelligence," *Journal of Experimental Psychology: Learning, Memory, and Cognition* 48, no. 5 (May 2022): 717–33, https://doi.org/10.1037/xlm0001135.

157 **One is having an active social network:** The Livingston et al., "Dementia Prevention" article describes the eleven factors affecting dementia prevention. And this general-audience piece discusses the benefits of friendship in elderly adults: Lydia Denworth, "Friendship Is a Lifesaver," *Nautilus*, April 15, 2020, https://nautil.us/friendship-is-a-lifesaver-237778/.

157 **Talking creates an important practice:** As mentioned in earlier chapters, recalling information strengthens long-term memory for it. Jeffrey D. Karpicke, "Retrieval-Based Learning: Active Retrieval Promotes Meaningful Learning," *Current Directions in Psychological Science* 21, no. 3 (June 2012): 157–63, https://doi.org/10.1177/0963721412443552.

157 **Self-talk does appear to help:** Jutta Kray, Jutta Eber, and Ulman Lindenberger, "Age Differences in Executive Functioning across the Lifespan: The Role of Verbalization in Task Preparation," *Acta Psychologica* 115, nos. 2–3 (February–March 2004): 143–65, https://doi.org/10.1016/j.actpsy.2003.12.001.

159 **In one study, college-student volunteers:** Barbara J. Reinke, David S. Holmes, and Nancy W. Denney, "Influence of a 'Friendly Visitor' Program on the Cognitive Functioning and Morale of Elderly Persons," *American Journal of Community Psychology* 9, no. 4 (August 1981): 491–504, https://onlinelibrary.wiley.com/doi/10.1007/BF00918178.

159 **The researchers tried a huge talking intervention:** Chang-Hoon Gong and Shinichi Sato, "Can Mild Cognitive Impairment with Depression Be Improved Merely by Exercises of Recall Memories Accompanying Everyday Conversation? A Longitudinal Study 2016–2019," *Quality in Ageing and*

Older Adults 23, no. 1 (January 2022): 26–35, https://doi.org/10.1108/QAOA-09-2021-0069.

161 **Some studies have used software:** Sandra Derbring et al., "Effects of a Digital Reminiscing Intervention on People with Dementia and Their Care-Givers and Relatives," *Ageing & Society* 43, no. 9 (September 2023): 1983–2000, https://doi.org/10.1017/S0144686X21001446.

161 **Another project connects elderly adults:** Hiroko H. Dodge et al., "Web-Enabled Conversational Interactions as a Method to Improve Cognitive Functions: Results of a 6-Week Randomized Controlled Trial," *Alzheimer's & Dementia: Translational Research & Clinical Interventions* 1, no. 1 (June 2015): 1–12, https://doi.org/10.1016/j.trci.2015.01.001.

161 **Conversational agents can be:** Christiane Even et al., "Benefits and Challenges of Conversational Agents in Older Adults," *Zeitschrift für Gerontologie und Geriatrie* 55, no. 5 (August 2022): 381–87, https://doi.org/10.1007/s00391-022-02085-9.

162 **Japan has invested:** This article describes Japan's experiences: James Wright, "Inside Japan's Long Experiment in Automating Elder Care," *MIT Technology Review*, January 9, 2023, https://www.technologyreview.com/2023/01/09/1065135/japan-automating-eldercare-robots/. And this article describes the promise of robots more generally: Neil Savage, "Robots Rise to Meet the Challenge of Caring for Old People," *Nature* 601, no. 7893 (January 19, 2022): S8–10, https://doi.org/10.1038/d41586-022-00072-z.

163 **often vocabulary increases:** Kavé, "Vocabulary Changes in Adulthood."

164 **Bernice and Roy could use some help:** Emily Sohn, "How Decision-Making Changes with Age," Simons Foundation, March 2, 2022, https://www.simonsfoundation.org/2022/03/02/how-decision-making-changes-with-age/.

165 **Card and board games:** Yao Ching-Teng, "Effect of Board Game Activities on Cognitive Function Improvement among Older Adults in Adult Day Care Centers," *Social Work in Health Care* 58, no. 9 (October 21, 2019): 825–38, https://doi.org/10.1080/00981389.2019.1656143.

165 **Reminiscence writing programs:** Kate de Medeiros et al., "The Impact of Autobiographic Writing on Memory Performance in Older Adults: A Prelim-

inary Investigation," *American Journal of Geriatric Psychiatry* 15, no. 3 (March 2007): 257–61, https://doi.org/10.1097/01.JGP.0000240985.10411.3e.

166 **Crossword puzzles put:** Mike Murphy, Katie O'Sullivan, and Kieran G. Kelleher, "Daily Crosswords Improve Verbal Fluency: A Brief Intervention Study," *International Journal of Geriatric Psychiatry* 29, no. 9 (September 2014): 915–19, https://doi.org/10.1002/gps.4079.

166 **Other word games:** This website offers a few examples: "Games & Quizzes," Merriam-Webster, accessed August 13, 2023, https://www.merriam-webster .com/games.

167 **Older adults tend to prioritize:** Laura L. Carstensen and Megan E. Reynolds, "Age Differences in Preferences through the Lens of Socioemotional Selectivity Theory," *Journal of the Economics of Ageing* 24 (February 2023): 100440, https://doi.org/10.1016/j.jeoa.2022.100440.

169 **Gerontologists do study the factors:** Manoj Pardasani, "Senior Centers: Characteristics of Participants and Nonparticipants," *Activities, Adaptation & Aging* 34, no. 1 (March 17, 2010): 48–70, https://doi.org/10.1080/01924780 903552295.

Chapter 7: Talking Changeth Language

173 **"Everybody say":** Cameo, "Word Up!," words and music by Larry Blackmon and Tomi Jenkins, Lyrics © Universal Music Publishing Group, 1986. I'm using these lyrics as a description of the pressures of talk planning; when it's time to talk, we've got to get going quickly and get our words up from long-term memory and into the talk-planning system. That's not the original meaning of these lyrics, but in a chapter including discussion on how talking increases ambiguity, I think that's just fine.

174 **This survival-of-the-fittest perspective:** Dan Dediu et al., "Cultural Evolution of Language," in *Cultural Evolution: Society, Technology, Language, and Religion*, ed. Peter J. Richerson and Morten H. Christiansen (MIT Press, 2013), https://direct.mit.edu/books/edited-volume/4020/chapter /167226/Cultural-Evolution-of-Language.

174 **When linguists think about:** Henry Brighton, Simon Kirby, and Kenny Smith, "Cultural Selection for Learnability: Three Principles Underlying the

View That Language Adapts to Be Learnable," in *Language Origins: Perspectives on Evolution*, ed. Maggie Tallerman (Oxford University Press, 2005), 291–309.

175 **We do the jaw open-close:** Peter F. MacNeilage, "The Frame/Content Theory of Evolution of Speech Production," *Behavioral and Brain Sciences* 21, no. 4 (August 1998): 499–511, https://doi.org/10.1017/S0140525X98001265.

175 **Hawai'ian, the language spoken:** This website is the University of Hawai'i at Hilo's guide to the Hawai'ian language: https://olelo.hawaii.edu/en/olelo.

176 **Dry air influences talking difficulty:** Caleb Everett, "Languages in Drier Climates Use Fewer Vowels," *Frontiers in Psychology* 8 (2017): 1285, https://www.frontiersin.org/articles/10.3389/fpsyg.2017.01285.

176 **Vowel sounds travel well through warm air:** Ian Maddieson, "Language Adapts to Environment: Sonority and Temperature," *Frontiers in Communication* 3 (2018): 28, https://doi.org/10.3389/fcomm.2018.00028.

177 **including the European starling:** "Common Starling Sounds, Starling Call," posted April 13, 2023, by Happy PonyTail, YouTube, https://www.youtube.com/watch?v=9zbx9VDzLf0; "White Bellbirds Produce Loudest Bird Call Ever Recorded," posted October 21, 2019, by SciNews, YouTube, https://www.youtube.com/watch?v=nREYBx4eZf8.

179 **Most words have several meanings:** Jennifer M. Rodd, "Settling into Semantic Space: An Ambiguity-Focused Account of Word-Meaning Access," *Perspectives on Psychological Science* 15, no. 2 (March 2020): 411–27, https://doi.org/10.1177/1745691619885860.

179 **When we're the audience:** Delphine Dahan and Gareth Gaskell, "The Temporal Dynamics of Ambiguity Resolution: Evidence from Spoken-Word Recognition," *Journal of Memory and Language* 57, no. 4 (November 2007): 483–501, https://doi.org/10.1016/j.jml.2007.01.001.

179 **People figure out a talker's:** Mark S. Seidenberg and Maryellen C. MacDonald, "The Impact of Language Experience on Language and Reading: A Statistical Learning Approach," *Topics in Language Disorders* 38, no. 1 (January/March 2018): 66–83, https://doi.org/10.1097/TLD.0000000000000144.

180 **As we're picking words:** Victor S. Ferreira, L. Robert Slevc, and Erin S. Rogers, "How Do Speakers Avoid Ambiguous Linguistic Expressions?," *Cognition* 96, no. 3 (July 2005): 263–84, https://doi.org/10.1016/j.cognition.2004.09.002.

180 **It's not an accident that common words:** Steven T. Piantadosi, Harry Tily, and Edward Gibson, "The Communicative Function of Ambiguity in Language," *Cognition* 122, no. 3 (March 2012): 280–91, https://doi.org/10.1016/j.cognition.2011.10.004.

182 **With three words to arrange:** Russell S. Tomlin, *Basic Word Order (RLE Linguistics B: Grammar): Functional Principles* (Routledge, 2014).

183 **One idea is that all human languages:** Murray Gell-Mann and Merritt Ruhlen, "The Origin and Evolution of Word Order," *Proceedings of the National Academy of Sciences of the United States of America* 108, no. 42 (October 18, 2011): 17290–95, https://doi.org/10.1073/pnas.1113716108.

184 **Most sign languages arose:** Connie de Vos and Roland Pfau, "Sign Language Typology: The Contribution of Rural Sign Languages," *Annual Review of Linguistics* 1, no. 1 (2015): 265–88, https://10.1146/annurev-linguist-030514-124958.

184 **A different kind of hypothesis:** Irit Meir et al., "The Effect of Being Human and the Basis of Grammatical Word Order: Insights from Novel Communication Systems and Young Sign Languages," *Cognition* 158 (January 2017): 189–207, https://doi.org/10.1016/j.cognition.2016.10.011; R. Jackendoff, *Foundations of Language: Brain, Meaning, Grammar, Evolution* (Oxford University Press, 2002).

184 **We certainly talk about ourselves:** Alessandra Macbeth et al., "Using the Electronically Activated Recorder (EAR) to Capture the Day-to-Day Linguistic Experiences of Young Adults," *Collabra: Psychology* 8, no. 1 (June 22, 2022): 36310, https://doi.org/10.1525/collabra.36310.

184 **A bias to mention ourselves:** Adele Goldberg, *Constructions at Work: The Nature of Generalization in Language* (Oxford University Press, 2006); Wallace Chafe, "Givenness, Contrastiveness, Definiteness, Subjects, Topics, and Point of View," in *Subject and Topic*, ed. Charles Li (Academic Press, 1976), 27–55.

185 **Writers and orators:** The idea of saving the best for last is incredibly widespread. Here's one example: "Writing for Success: Argument," QuillBot, accessed June 19, 2022, https://courses.lumenlearning.com/englishcomp1v2xmaster/chapter/writing-for-success-argument/.

185 **Memory research shows:** Julieta Laurino and Laura Kaczer, "Animacy as a Memory Enhancer during Novel Word Learning: Evidence from Orthographic and Semantic Memory Tasks," *Memory* 27, no. 6 (July 3, 2019): 820–28, https://doi.org/10.1080/09658211.2019.1572195.

185 **Easy-first is the bias:** J. Kathryn Bock, "Toward a Cognitive Psychology of Syntax: Information Processing Contributions to Sentence Formulation," *Psychological Review* 89, no. 1 (January 1982): 1–47, https://doi.org/10.1037/0033-295X.89.1.1.

187 **Writing style guides concur:** Here are a few of the many recommendations to avoid passive sentences: "1. Why Do English Teachers Hate the Passive Voice? Okay, So . . .," CliffsNotes, accessed January 22, 2023, https://www.cliffsnotes.com/tutors-problems/Writing/48522220-1-Why-do-English-teachers-hate-the-passive-voice-Okay-so/; "Avoiding Passive Voice," *Concentric*, June 6, 2023, https://www.concentric.io/blog/avoiding-passive-voice; Christopher Taylor et al., "How to Avoid Using the Passive Voice," WikiHow, September 30, 2022, https://www.wikihow.com/Avoid-Using-the-Passive-Voice.

189 **Researchers used simple subliminal images:** Lila R. Gleitman et al., "On the *Give* and *Take* between Event Apprehension and Utterance Formulation," *Journal of Memory and Language* 57, no. 4 (November 2007): 544–69, https://doi.org/10.1016/j.jml.2007.01.007.

190 **The film was based:** *The Wizard of Oz*, directed by Victor Fleming and King Vidor (1939; Metro-Goldwyn-Mayer). The novel—L. Frank Baum, *The Wonderful Wizard of Oz* (Geo. M. Hill Co., 1900), https://www.loc.gov/resource/rbc0001.2006gen32405/—is available from Project Gutenberg: https://www.gutenberg.org/ebooks/55.

191 **Another reason *yellow-brick road*:** Christine Günther, "A Difficult to Explain Phenomenon: Increasing Complexity in the Prenominal Position," *English Language & Linguistics* 23, no. 3 (September 2019): 645–70, https://doi.org/10.1017/S1360674318000084.

192 **Consider our QWERTY keyboard:** Here's a Dvorak keyboard layout: "Dvorak Keyboard Layout," Wikipedia, accessed July 23, 2023, https://en.wikipedia.org/wiki/Dvorak_keyboard_layout. On resistance to the metric system: M. Wober, "Attitudes towards Metric and Imperial Systems of Measurement," in *Language*, ed. Howard Giles, W. Peter Robinson, and Philip M. Smith (Pergamon, 1980), 223–26, https://doi.org/10.1016/B978-0-08-024696-3.50038-7.

192 **The language is called Speedtalk:** Robert A. Heinlein's 1949 novella *Gulf* is available here: https://www.baen.com/Chapters/9781451637854B/9781451637854B___4.htm.

193 **Speedtalk hits the trifecta:** Here's one study about learning very similar short words: James S. Magnuson et al., "The Time Course of Spoken Word Learning and Recognition: Studies with Artificial Lexicons," *Journal of Experimental Psychology: General* 132, no. 2 (June 2003): 202–27, https://doi.org/10.1037/0096-3445.132.2.202.

194 **Having thousands of very similar:** Daniel J. Acheson and Maryellen C. MacDonald, "Twisting Tongues and Memories: Explorations of the Relationship between Language Production and Verbal Working Memory," *Journal of Memory and Language* 60, no. 3 (April 2009): 329–50, https://doi.org/10.1016/j.jml.2008.12.002.

194 **"Everything is more complicated":** Hope Jahren, *Lab Girl: A Memoir* (Knopf Doubleday, 2016), 29.

Chapter 8: By Your Talking You Shall Be Judged

195 **"Zora laughed":** Zadie Smith, *On Beauty* (Thorndike Press, 2006).

196 **Some folks who want to leave:** This blog post discusses various accent reduction programs: Emma Miller, "Top Accent Reduction Classes in the USA [2023]," *Medium* (blog), March 5, 2024, https://medium.com/@emmamillerw1990/top-accent-reduction-class-in-the-usa-2023-de3d8dc54f2d.

197 **Bilinguals, by virtue:** Salim Abu-Rabia and Ekaterina Sanitsky, "Advantages of Bilinguals over Monolinguals in Learning a Third Language," *Bilingual Research Journal* 33, no. 2 (August 25, 2010): 173–99, https://doi.org/10.1080/15235882.2010.502797.

199 **sign language accents crop up:** Wendy Sandler, Gal Belsitzman, and Irit Meir, "Visual Foreign Accent in an Emerging Sign Language," *Sign Language & Linguistics* 23, nos. 1–2 (October 2020): 233–57, https://doi.org/10.1075/sll.00050.san.

199 **Dialect includes:** This quiz mixes questions about accent and word choices to attempt to guess someone's US English dialect: Josh Katz and Wilson Andrews, "How Y'all, Youse and You Guys Talk," *New York Times*, December 21, 2013, https://www.nytimes.com/interactive/2014/upshot/dialect-quiz-map.html It's fun even if you're not a native speaker of English or not from the United States.

199 **Children are actively learning and reproducing:** Aurélie Nardy, Jean-Pierre Chevrot, and Stéphanie Barbu, "Sociolinguistic Convergence and Social Interactions within a Group of Preschoolers: A Longitudinal Study," *Language Variation and Change* 26, no. 3 (October 2014): 273–301, https://doi.org/10.1017/S0954394514000131.

200 **Whatever the nature of this prejudice:** This article reviews many studies of kids' accent and dialect biases: Kana Imuta and Jessica L. Spence, "Developments in the Social Meaning Underlying Accent- and Dialect-Based Social Preferences," *Child Development Perspectives* 14, no. 3 (September 2020): 135–41, https://doi.org/10.1111/cdep.12374.

200 **And when asked to choose:** Katherine D. Kinzler et al., "Accent Trumps Race in Guiding Children's Social Preferences," *Social Cognition* 27, no. 4 (August 2009): 623–34, https://doi.org/10.1521/soco.2009.27.4.623.

201 **These childhood biases:** Melissa Paquette-Smith et al., "The Effect of Accent Exposure on Children's Sociolinguistic Evaluation of Peers," *Developmental Psychology* 55, no. 4 (2019): 809–22, https://doi.org/10.1037/dev0000659.

201 **These accent biases just keep coming:** An example of a new journal on this topic is the *Journal of Language Discrimination*, https://journal.equinoxpub.com/JLD.

201 **In the 1980s, Baugh accepted:** This podcast (with printed transcript available) is an interview with John Baugh about his initial experiences with hous-

ing discrimination and how he set out to research the problem: "ASHA Voices: The Effects of Linguistic Profiling," ASHAWire, accessed September 20, 2023, https://leader.pubs.asha.org/do/10.1044/2022-1013-transvoices-dialect-baugh-linguistics.

203 **Since Baugh's initial study:** Justin T. Craft et al., "Language and Discrimination: Generating Meaning, Perceiving Identities, and Discriminating Outcomes," *Annual Review of Linguistics* 6 (January 14, 2020): 389–407, https://doi.org/10.1146/annurev-linguistics-011718-011659.

204 **Many societies signal respect:** This page describes some cultural variation in funeral attire: "Funeral Rites across Different Cultures: The Significance of Color," Woodlawn, accessed August 13, 2023, https://www.woodlawn.org/blog/funeral-rites-across-different-cultures-the-significance-of-color/.

205 **English is often described as having:** "English: 3 Distinctly Different Dialects That Are Spoken in the United States," Atlas, February 6, 2024, https://atlasls.com/english-3-different-dialects-spoken-united-states/.

205 **Yes, all dialects are equally good:** This PBS site explains many dialect facts, focusing on American English dialects: "Do You Speak American?," PBS, accessed March 29, 2023, https://www.pbs.org/speak/seatosea/american varieties/.

205 **The linguist Lisa Davidson:** Her comments about prejudice against talking styles were in this interview, which also provides a good review of the uproar surrounding vocal fry, described later in this chapter: Eileen Reynolds, "What's the Big Deal about Vocal Fry? An NYU Linguist Weighs In," NYU, September 29, 2015, https://www.nyu.edu/about/news-publications/news/2015/september/lisa-davidson-on-vocal-fry.html.

206 **Around 60 to 80 percent of Americans:** Laura Wagner et al., "To What Extent Does the General Public Endorse Language Myths?," *Language and Linguistics Compass* 17, no. 3 (May/June 2023): e12486, https://doi.org/10.1111/lnc3.12486.

207 **Some 2,500 years ago:** "Sanskrit Language," Britannica, accessed April 19, 2023, https://www.britannica.com/topic/Sanskrit-language.

208 ***Hello* was once considered:** Some of the history of *hello*: Robert Krulwich,

"A (Shockingly) Short History of 'Hello,'" NPR, February 17, 2011, https://www.npr.org/sections/krulwich/2011/02/17/133785829/a-shockingly-short-history-of-hello.

209 **The level of hostility and alarm:** This first article suggests that vocal fry leads to women being perceived as less competent: Sally K. Gallena and James A. Pinto, "How Graduate Students with Vocal Fry Are Perceived by Speech-Language Pathologists," *Perspectives of the ASHA Special Interest Groups* 6, no. 6 (December 2021): 1554–65, https://doi.org/10.1044/2021_PERSP-21-00083. And this second one calls out the problems with that approach: Matthew B. Winn, Alayo Tripp, and Benjamin Munson, "A Critique and Call for Action, in Response to Sexist Commentary about Vocal Fry," *Perspectives of the ASHA Special Interest Groups* 7, no. 6 (December 2022): 1903–7, https://doi.org/10.1044/2022_PERSP-21-00319.

209 **The etiquette columnist Miss Manners:** "Please Change Your Voice So It Doesn't Annoy Me," Miss Manners, UExpress, November 16, 2019, https://www.uexpress.com/life/miss-manners/2019/11/16.

210 **In previous decades, vocal fry:** This article reviews much of the history of vocal fry: Nassima B. Abdelli-Beruh, Lesley Wolk, and Dianne Slavin, "Prevalence of Vocal Fry in Young Adult Male American English Speakers," *Journal of Voice* 28, no. 2 (2014): 185–90, https://doi.org/10.1016/j.jvoice.2013.08.011.

210 **The bias against vocal fry:** This video from *The Wall Street Journal* gives examples of vocal fry and describes research about perceptions of men's and women's vocal fry: Francesca Fontana and Denise Blostein, "Young Women Speak, Older Ears Hear Vocal Fry," *Wall Street Journal*, October 19, 2017, https://www.wsj.com/articles/young-women-speak-older-ears-hear-vocal-fry-1507649162 .

210 **As a result, young women:** Amanda Hess in *Slate* describes these and other biases against vocal fry: Amanda Hess, "Why Old Men Find Young Women's Voices So Annoying," *Slate*, January 7, 2013, https://slate.com/human-interest/2013/01/vocal-fry-and-valley-girls-why-old-men-find-young-women-s-voices-so-annoying.html.

211 **The linguist Alexandra D'Arcy:** Her comments about women being judged for their talk were made on the *Vocal Fries* podcast, which includes a transcript here: "Learning to Love Like" transcript, *Vocal Fries Pod*, July 21, 2021, https://vocalfriespod.com/2021/07/21/learning-to-love-like-transcript/.

211 **Uptalk goes in the opposite direction:** This book covers many aspects of uptalk: Paul Warren, *Uptalk: The Phenomenon of Rising Intonation* (Cambridge University Press, 2016).

211 **It has even been observed in Spanish:** Ji Young Kim, "Spanish–English Cross-Linguistic Influence on Heritage Bilinguals' Production of Uptalk," *Languages* 8, no. 1 (March 2023): 22, https://doi.org/10.3390/languages8010022.

212 **Perhaps uptalk is especially useful:** Adrienne B. Hancock and Benjamin A. Rubin, "Influence of Communication Partner's Gender on Language," *Journal of Language and Social Psychology* 34, no. 1 (January 2015): 46–64, https://doi.org/10.1177/0261927X14533197.

212 ***Like* has many different uses:** Alexandra D'Arcy, *Discourse-Pragmatic Variation in Context: Eight Hundred Years of LIKE*, Studies in Language Companion Series 187 (John Benjamins, 2017), https://benjamins.com/catalog/slcs.187.

213 **Some versions of these *likes* can be found:** D'Arcy, *Discourse-Pragmatic Variation in Context.*

213 **A study of college students' speech:** Alessandra Macbeth et al., "Using the Electronically Activated Recorder (EAR) to Capture the Day-to-Day Linguistic Experiences of Young Adults," *Collabra: Psychology* 8, no. 1 (June 22, 2022): 36310, https://doi.org/10.1525/collabra.36310.

215 **For example, the suggestion:** Ben Zimmer's article on *Ms.* traces much of its history: Ben Zimmer, "Hunting the Elusive First 'Ms.,'" Visual Thesaurus, June 23, 2009, https://www.visualthesaurus.com/cm/wordroutes/hunting-the-elusive-first-ms/.

216 **Like *Ms.*, gender-neutral pronouns:** Dennis Baron, *What's Your Pronoun? Beyond He and She* (Liveright, 2020).

216 **One in five Americans:** A. W. Geiger and Nikki Graf, "About One-in-Five U.S. Adults Know Someone Who Goes by a Gender-Neutral Pronoun," Pew Research Center, September 5, 2019, https://www.pewresearch.org/short-reads/2019/09/05/gender-neutral-pronouns/.

216 **Given that talking is harder than comprehending:** Dennis Baron, "Pronoun Backlash," *The Web of Language* (blog), October 4, 2020, https://blogs.illinois.edu/view/25/49786652.

218 **This partial deviation in the verb:** This article describes children's acquisition of verb tenses over time: Mabel L. Rice, Kenneth Wexler, and Scott Hershberger, "Tense over Time," *Journal of Speech, Language, and Hearing Research* 41, no. 6 (December 1998): 1412–31, https://doi.org/10.1044/jslhr.4106.1412.

219 **I've included some suggestions:** This is the first of several helpful blogs by a linguist: Kirby Conrod, "Pronouns 101: Introduction to Your Loved One's New Pronouns," *Medium* (blog), December 27, 2020, https://kconrod.medium.com/pronouns-101-introduction-to-your-loved-ones-new-pronouns-3fef080266d0. This post from the Mayo Clinic website also offers good advice, with additional links in the post: Cesar Gonzalez, "How Can I Train Myself to Use the Right Pronouns?," *Mayo Clinic Press* (blog), December 7, 2021, https://mcpress.mayoclinic.org/living-well/how-can-i-train-myself-to-use-the-right-pronouns/.

219 **Some examples include the rejection:** Here are two of many different perspectives on potentially offensive word choices and their alternatives. John McWhorter, *Nine Nasty Words: English in the Gutter: Then, Now, and Forever* (Penguin, 2021); George Packer, "The Moral Case against Equity Language," *The Atlantic*, March 2, 2023, https://www.theatlantic.com/magazine/archive/2023/04/equity-language-guides-sierra-club-banned-words/673085/.

220 **Words can be fists:** Sigrid Nunez, *The Friend* (Riverhead Books, 2018).

Chapter 9: The Science of Talk Analysis

221 **"Black words on a white page":** This Guy de Maupassant quote pops up everywhere, but never with an attribution. It's probably in a letter.

222 **Experts in overtalking estimate:** Here is a test that researchers use to measure overtalking: "Talkaholic Score," accessed April 14, 2023, http://www.jamescmccroskey.com/measures/compulsive_communication.htm. And here are two articles about navigating a conversation or relationship with an overtalker, which may also provide some insight for overtalkers themselves: F. Diane Barth, "5 Steps for Dealing with People Who Talk Too Much," *Psychology Today*, April 22, 2012, https://www.psychologytoday.com/us/blog/the-couch/201204/5-steps-dealing-people-who-talk-too-much; Nancy Wartik, "What to Do about an Overtalker," *New York Times*, updated December 21, 2019, https://www.nytimes.com/2019/12/11/well/what-to-do-about-an-overtalker.html.

223 **In the absence of extensive shared:** Nicolas Fay et al., "Socially Situated Transmission: The Bias to Transmit Negative Information Is Moderated by the Social Context," *Cognitive Science* 45, no. 9 (September 2021): e13033, https://doi.org/10.1111/cogs.13033.

224 **In one episode:** *Dexter*, season 2, episode 7, "That Night, a Forest Grew," written by Daniel Cerone, directed by Jeremy Podeswa, aired November 11, 2007, on Showtime.

226 **Iris Murdoch was an illustrious British philosopher:** Peter Garrard et al., "The Effects of Very Early Alzheimer's Disease on the Characteristics of Writing by a Renowned Author," *Brain* 128, no. 2 (February 2005): 250–60, https://doi.org/10.1093/brain/awh341.

226 **Reagan was diagnosed:** Visar Berisha et al., "Tracking Discourse Complexity Preceding Alzheimer's Disease Diagnosis: A Case Study Comparing the Press Conferences of Presidents Ronald Reagan and George Herbert Walker Bush," *Journal of Alzheimer's Disease* 45, no. 3 (January 2015): 959–63, https://doi.org/10.3233/JAD-142763.

227 **About one in five adults:** Statistics about mental illness can be found here: "Mental Illness," National Institute of Mental Health, accessed July 16, 2023, https://www.nimh.nih.gov/health/statistics/mental-illness.

229 **Words that carry a lot of meaning:** Eduardo G. Altmann, Janet B. Pierrehumbert, and Adilson E. Motter, "Beyond Word Frequency: Bursts, Lulls, and Scaling in the Temporal Distributions of Words," *PLOS One* 4, no. 11

(November 11, 2009): e7678, https://doi.org/10.1371/journal.pone.0007678.

230 **Compared to people with no mental illness:** This article discusses both pronouns and negative-emotion words in depression: Allison M. Tackman et al., "Depression, Negative Emotionality, and Self-Referential Language: A Multi-Lab, Multi-Measure, and Multi-Language-Task Research Synthesis," *Journal of Personality and Social Psychology* 116, no. 5 (May 2019): 817–34, https://doi.org/10.1037/pspp0000187.

231 **As I've mentioned before:** Alessandra Macbeth et al., "Using the Electronically Activated Recorder (EAR) to Capture the Day-to-Day Linguistic Experiences of Young Adults," *Collabra: Psychology* 8, no. 1 (June 22, 2022): 36310, https://doi.org/10.1525/collabra.36310. We don't know the mental health status of the participants in this study, but we can infer from other statistics that probably most of them did not have a mental illness diagnosis, meaning that self-pronoun use among people without mental health challenges is also very high.

231 **Putting self-pronouns and thinking words together:** This article reviews these and other links between mental illness and talking: Charlotte Entwistle, Ely Marceau, and Ryan L. Boyd, "Personality Disorder and Verbal Behavior," in *Handbook of Language Analysis in Psychology*, ed. Morteza Dehghani and Ryan L. Boyd (Guilford Press, 2022), https://eprints.lancs.ac.uk/id/eprint/154939/.

232 **The interviewer, the actor Sean Penn:** Sean Penn's interview with Charles Bukowski dates from September 1987 in *Interview* magazine. Bukowski's comments in the interview are reprinted here: "Tough Guys Write Poetry: Charles Bukowski Interviewed by Sean Penn," *Southern Cross Review*, https://southerncrossreview.org/97/bukowski-penn.htm.

233 **Research on the safety:** Nduma N. Basil et al., "Health Records Database and Inherent Security Concerns: A Review of the Literature," *Cureus* 14, no. 10 (2022): e30168, https://doi.org/10.7759/cureus.30168.

234 **Most personality researchers think:** Wiebke Bleidorn et al., "Personality Stability and Change: A Meta-Analysis of Longitudinal Studies," *Psycholog-*

ical Bulletin 148, nos. 7–8 (2022): 588–619, https://doi.org/10.1037/bul0000365.

235 **All of these Facebook users:** Gregory Park et al., "Automatic Personality Assessment through Social Media Language," *Journal of Personality and Social Psychology* 108, no. 6 (June 2015): 934–52, https://doi.org/10.1037/pspp0000020.

239 **Left to our own intuitions:** Daria Plotkina, Andreas Munzel, and Jessie Pallud, "Illusions of Truth—Experimental Insights into Human and Algorithmic Detections of Fake Online Reviews," *Journal of Business Research* 109 (March 2020): 511–23, https://doi.org/10.1016/j.jbusres.2018.12.009.

239 **Fake news and other lies:** This article from Snopes lists many medical myths and conspiracy theories: "Coronavirus Is a Breeding Ground for Conspiracy Theories," Snopes, February 28, 2020, https://www.snopes.com/news/2020/02/28/coronavirus-is-a-breeding-ground-for-conspiracy-theories/. And this Snopes article lists many money-related scams: "List: 24 Online Scams to Watch Out For," Snopes, February 22, 2022, https://www.snopes.com/collections/online-scam-collection/.

240 **Several education programs:** Ullrich K. H. Ecker et al., "The Psychological Drivers of Misinformation Belief and Its Resistance to Correction," *Nature Reviews Psychology* 1, no. 1 (January 2022): 13–29, https://doi.org/10.1038/s44159-021-00006-y.

240 **The first device to offer:** John Synnott, David Dietzel, and Maria Ioannou, "A Review of the Polygraph: History, Methodology and Current Status," *Crime Psychology Review* 1, no. 1 (January 2015): 59–83, https://doi.org/10.1080/23744006.2015.1060080.

241 **Freud believed that:** Gary Dell, "Speaking and Misspeaking," in *An Invitation to Cognitive Science, Volume 1: Language*, ed. Lila Gleitman, Mark Liberman, and Daniel Osherson (MIT Press, 1995), https://doi.org/10.7551/mitpress/3964.001.0001.

242 **The liar's extra difficulty:** Matthew L. Newman et al., "Lying Words: Predicting Deception from Linguistic Styles," *Personality and Social Psychology*

Bulletin 29, no. 5 (May 1, 2003): 665–75, https://doi.org/10.1177/014616720 3029005010. This study is a source both for poor human judgments and the effects of lying on talking.

244 **The headlines below:** Gordon Pennycook et al., "A Practical Guide to Doing Behavioral Research on Fake News and Misinformation," *Collabra: Psychology* 7, no. 1 (July 13, 2021): 25293, https://doi.org/10.1525/collabra.25293.

245 **but fake news is designed:** This article from *Wired* magazine reviews properties of fake news (surprising, emotional, etc.) and offers some suggestions for trying to distinguish real from fake: David Nield, "4 Tips to Spot Misinformation on the Web," *Wired*, November 26, 2023, https://www.wired.com/story/4-tips-spot-misinformation/.

245 **A study of fake news on X:** Robinson Meyer, "The Grim Conclusions of the Largest-Ever Study of Fake News," *The Atlantic*, March 8, 2018, https://www.theatlantic.com/technology/archive/2018/03/largest-study-ever-fake-news-mit-twitter/555104/.

246 **Good detection of fake news requires:** Xinyi Zhou and Reza Zafarani, "A Survey of Fake News: Fundamental Theories, Detection Methods, and Opportunities," *ACM Computing Surveys* 53, no. 5 (September 2021): 1–40, https://doi.org/10.1145/3395046.

247 **folks who are more analytic:** Gordon Pennycook and David G. Rand, "Lazy, Not Biased: Susceptibility to Partisan Fake News Is Better Explained by Lack of Reasoning Than by Motivated Reasoning," *Cognition* 188 (July 2019): 39–50, https://doi.org/10.1016/j.cognition.2018.06.011.

247 **Here is young Harriet Smith:** Jane Austen, *Emma* (R. Bentley, 1886), available from Project Gutenberg, https://www.gutenberg.org/ebooks/158.

249 **A majority of Americans believe:** Michelle Faverio, "Key Findings about Americans and Data Privacy," Pew Research Center, October 18, 2023, https://www.pewresearch.org/short-reads/2023/10/18/key-findings-about-americans-and-data-privacy/.

Afterword

251 **There's a common piece of writing advice:** After seeing this advice given both without attribution and with attribution to Gogol, I asked an expert in nineteenth-century Russian literature, Maksim Hanukai, whether there is any evidence that Nikolai Gogol said this. Maksim reported that he could find no evidence that this advice originated with Gogol.

255 **A colleague who studies video games:** That colleague is C. Shawn Green, https://psych.wisc.edu/staff/green-c-shawn/.

INDEX

INDEX